CAMBRIDGE LIBRARY COLLECTION

Books of enduring scholarly value

Technology

The focus of this series is engineering, broadly construed. It covers technological innovation from a range of periods and cultures, but centres on the
technological achievements of the industrial era in the West, particularly in
the nineteenth century, as understood by their contemporaries. Infrastructure
is one major focus, covering the building of railways and canals, bridges and
tunnels, land drainage, the laying of submarine cables, and the construction
of docks and lighthouses. Other key topics include developments in industrial
and manufacturing fields such as mining technology, the production of iron
and steel, the use of steam power, and chemical processes such as photography
and textile dyes.

The Life of Sir John Fowler, Engineer

As a civil engineer, Sir John Fowler (1817–98) devoted his life to the railways.
His best-known achievements include the first railway bridge across the Thames
in London, Manchester Central Station, the development of the London
Underground and (with Sir Benjamin Baker) the Forth Bridge – arguably the
most remarkable feat of engineering of the nineteenth century. Given access
to friends and family papers, the author and social theorist Thomas Mackay
(1849–1912) portrays a man who was fascinated by engineering as a child,
and who continued to work up until his death. As a portrait of one of the
architects of Victorian Britain, this biography, first published in 1900, will be
of great interest to historians of the period as well as readers wishing to know
more about the development of iconic infrastructure.

T0188064

The Life of Sir John Fowler, Engineer

Thomas Mackay

CAMBRIDGE
UNIVERSITY PRESS

CAMBRIDGE UNIVERSITY PRESS

Cambridge, New York, Melbourne, Madrid, Cape Town,
Singapore, São Paolo, Delhi, Mexico City

Published in the United States of America by Cambridge University Press, New York

www.cambridge.org
Information on this title: www.cambridge.org/9781108057677

© in this compilation Cambridge University Press 2013

This edition first published 1900
This digitally printed version 2013

ISBN 978-1-108-05767-7 Paperback

Photo Elliott & Fry

THE LIFE OF
SIR JOHN FOWLER

ENGINEER

BART., K.C.M.G., Etc.

By THOMAS MACKAY

WITH EIGHTEEN ILLUSTRATIONS

LONDON
JOHN MURRAY, ALBEMARLE STREET
1900

PREFACE

THE life of a great practical engineer, whose labours cover the last sixty years of the nineteenth century, might without irrelevance be made the occasion of writing a history of modern engineering. It was Sir John Fowler's business to utilise, in the interest of his clients, the inventions and discoveries of a most prolific half-century of progress, and an enumeration of these would in itself constitute a very complete chronicle of the advance of engineering science.

In such a treatment of the subject the vastness of the details must have overwhelmed the features of any single personality, however eminent, and no such ambitious task has here been attempted. It has been the author's endeavour to sketch the lights and shades of a strong and interesting character, and to indicate the particular subdivision of scientific function to which Sir John Fowler's energies were so successfully devoted.

A great organiser, like Sir John Fowler, appeals perhaps less strongly to the imagination than the great discoverer or inventor. Still, his work is an indispensable element in that most important task—the domestication of science for the public service. Our conception of the industrial and commercial mechanism of the age will be very incomplete, unless we realise the part which is played by men like the subject of this biography.

A large collection of letters and papers has been placed in the author's hands by Lady Fowler. To her and to other members of the late Sir John Fowler's family he is much indebted for information, suggestions, and corrections. He has also to acknowledge valuable assistance given by the late Mr. Baldry, and by Sir Benjamin Baker, Sir John Fowler's partners. The kindness with which all his inquiries have been met has, he hopes, enabled him to overcome, in some degree, two great disadvantages : one that he did not know Sir John Fowler personally, the other that he is not an engineer.

The work of preparing the following narrative has been full of interest. An experience such as that of Sir John Fowler is a most important chapter of industrial history, and the author can only hope that in his presentation of the facts the interest has not been allowed entirely to evaporate.

CONTENTS

CHAPTER XI.

THE FORTH BRIDGE

CHAPTER XII.

THE ENGINEER AT HOME

APPENDIX

LIST OF ILLUSTRATIONS

LIFE

OF

SIR JOHN FOWLER

CHAPTER I.

EARLY LIFE

JOHN FOWLER, the subject of this biography, was born at his father's house, Wadsley Hall, near Sheffield, on the 15th of July, 1817. For more than 200 years the Fowler family had been connected with Wincobank, a district called after the hill of that name in the neighbourhood of Sheffield. John Fowler the father, himself the son of a John Fowler of Wincobank, was brought to be baptised at Ecclesfield Church on the 18th of June, 1784. Though living all his life in the neighbourhood of Sheffield, he never became closely connected with the mineral industries of the district. In early life he was among the young patriots of the day who took an active part in forming a regiment of volunteers to resist the French invasion threatened by Napoleon. At the age of nineteen he was lieutenant and quartermaster in a band of 200 raised in the neighbourhood of Ecclesfield. A false alarm, caused by the accidental lighting of a beacon

B

fire, called out the regiment and marched them off to
the east coast. This military incident in an otherwise
uneventful life was often the subject of some pleasant
merriment in the family circle, and as late as 1868 the
old gentleman wrote a brief account of the bloodless
career of the Ecclesfield volunteers, which Dr. Gatty
has incorporated in his edition of Hunter's *History of
Hallamshire*.

For five years Mr. Fowler gave close attention to his
military duties; he then retired and applied himself
to his profession, that of a land surveyor. It is
pleasant to think that within sound of the whirr
and grinding of the Sheffield machinery it was possible for one whose tastes lay in that direction to
win competence and reputation from the ancient
industry of the land, and to rear a family of healthy
and stalwart sons ready to play their part in the careers
which economic changes rendered for them inevitable.
His residence of Wadsley Hall was the place where
the courts of the manor of Wadsley were held, a link
with the vanishing order of things, yet standing on the
confines of the new industry, and soon to be the birth-
place of one who played an important part in the
economic revolution which was then pending. The
elder Fowler could resist the attraction which has ever
been drawing the population from the country and its
pursuits to the town, and to the triumphs, the sufferings, and the still unsolved problems of the modern
industrial system. John Fowler the son, as we shall
presently see, eagerly and confidently and with a real
enjoyment of work and struggle, threw himself into
the stream of the new industry, and was carried to
successes typical not only of the noble profession which

he had embraced, but indicative also of high personal qualities of courage, endurance, and skill.

To return, however, to his parentage. John Fowler the elder married on the 24th of December, 1815, Miss Elizabeth Swann, daughter of William Swann of Dykes Hall. In a letter, written at the time of his wife's death, to his daughter-in-law (wife of Sir John Fowler), Mr. Fowler recounts the history of his two years' courtship and of his marriage on Christmas Eve to Miss Swann; he relates how they decided to remain in his home at Wadsley instead of spending the honeymoon in travel, and how, in the drawing-room of the bride's new home, after the wedding guests had gone, he offered a prayer, specially prepared by himself for the occasion. This prayer found its answer in many long years of happy married life.

Mrs. Fowler belonged to the generation which valued highly the arts of the good housewife, and in these she excelled. To her intelligence and capacity were due the admirable arrangements for the comfort of the large family circle at Wadsley Hall. Her husband, who inherited from his father a fine physique and a stature of six feet, was a man of amiable and most trustworthy character. So highly esteemed was his probity, that he not infrequently was asked to act as sole valuator by contending parties. His habits of life were most methodical; up to the last year of his life he never retired to rest without making a personal inspection of his farm premises, in order to see that every animal had received its proper share of attention.

To this couple was born on the 15th of July, 1817, the subject of this biography, John Fowler the engineer. Other children came in due course:

William, who established large ironworks near Chesterfield; Henry, an engineer, who died on his return from
India, comparatively young; Charles, an architect, who
emigrated to Australia; Robert, a solicitor in Westminster; and Frederick, who succeeded to and extended the paternal business; one daughter married
Mr. Whitton, government engineer of railways in New
South Wales; a second married Captain Holmes, of
Norfolk; a third remained unmarried in attendance on
her father. Mrs. Fowler, the mother, died on the
5th of June, 1858. Her husband survived her for
many years, and died on the 19th of August, 1872, in
his eighty-ninth year.*

Mr. Fowler the elder was an energetic man of
business, but his interests lay somewhat apart from the
new industry. The career of John Fowler the son was
mapped out by himself. His early letters to his father,
though testifying to his filial affection and respect,
rarely ask for advice, but exhibit a confident, self-
reliant tone which is not a little remarkable in one so
young. The son, soon immersed in the restless whirl
of the new industry, is fond of poking a little kindly

---

* The Rev. J. EASTWOOD, in his learned *History of the Parish of Ecclesfield* (1862), gives some additional genealogical details. "Wadsley Hall,"
he says, "was rebuilt in 1722 by Charles Burton; it has been for the last
forty-seven years in the occupation of John Fowler, Esq., whose family
sprang from Wincobank in this parish. Joshua Fowler (born at Wincobank and baptised at Ecclesfield in 1678) died April 27th, 1742. His son
Samuel married, November 11th, 1731, Hannah, daughter of William
Dixon, of Shiregreen, and died April 6th, 1760. John, the son of Samuel,
born at Wincobank in 1746, married Hannah Webster (who died January
25th, 1829, aged seventy-four), and died May 25th, 1808. Their son was
the present Mr. Fowler, whose wife Elizabeth, daughter of William Swann
of Dykes Hall, died June 5th, 1858, aged sixty-six, leaving a numerous
family, the eldest of whom is John Fowler of Queen Square Place, Westminster, a distinguished engineer," etc., etc. (p. 448).

fun at the stay-at-home instincts of his parents, coupled with genuine expressions of regret that they cannot summon up energy to visit him at some of the temporary halting-places to which his professional engagements called him. The correspondence between father and son was continued to the end of the former's long life. There will be occasion to quote from it in the course of this narrative; one point, however, may here be noticed.

It was a characteristic of the younger man that he always sought to make men talk on their own subjects. In his letters to his father, more especially in later years, when his professional engagements carried him to the ends of the earth—to Spain, to Egypt, and to India—his familiar talk (for the family letter is a species of familiar talk) to his father is not of bridges and railways and hydraulic power, but of cattle and horses and agriculture.

This trait may perhaps appear again. Here it has only been noticed to mark by the son's testimony what we believe to have been the principal secular interest in his father's life.

The parish of Ecclesfield is for the most part a bleak moorland country, now much encroached upon by the advancing town. Roger Dodsworth, the Yorkshire topographer, visited Ecclesfield Church in 1628, and describes it as follows :—

"This church is called and that deservedly by the vulgar the Mynster of the Moores, being the fairest church for stone, wood, glass and neat keeping that ever I came in of country churches."

To this church, for many years after their marriage, Mr. and Mrs. Fowler used to ride across the fields

Sunday after Sunday, mounted on an old-fashioned pillion saddle. John Fowler, their eldest child, acquired by inheritance the vigour of a family reared for generations in this upland country. The Yorkshireman of the moors has the hardiness of his surroundings, but he lives too near the stir of the great world to entertain for long any of the dreamy listlessness often characteristic of those who, like the Highlander, are entirely secluded from the modern spirit. The family were in easy though not affluent circumstances. There was no question of the future engineer being made an eldest son, and so deprived of the inspiriting responsibility of earning his own living. At the same time he was not stinted of the sound but not ornamental education which was then available for the middle and professional classes of the country.

At the age of nine John Fowler was sent to Whitley Hall, a private school near Ecclesfield, kept by Mr. Rider. Mr. Rider is described in Mr. Eastwood's book as one " under whose gentle but efficient guidance many of the principal young men of the neighbourhood received their education."

With regard to his school days, Sir John Fowler in late life wrote down the following reminiscence, which is best given in his own words :—

"I remember two incidents there—one, being teased by an elder boy until a fight took place, when the elder boy had two front teeth knocked out. The mother of the boy was sent for, and a scene and examination occurred, resulting in my acquittal of all blame and my resolution never to fight again—a resolution I kept through life.

"Another incident was a fall from a wall and a cut on the eyebrows, leaving a well-defined cross, ever afterwards to be visible, by which I could be identified.

"At this early age I began a habit of telling stories in bed, which I invented in the dormitory where there were several boys, and gradually acquired such proficiency in making them more and more horrible until timid boys sometimes wept. Unfortunately one of the masters happened to hear something going on, and listened, and heard what induced him to put a stop to my improvising, or I might have acquired some curious proficiency.

"I shall never forget an attempt to frighten me by a big boy coming into the bedroom one night wrapped in a sheet, with a large hollow turnip on the top of his head and a lighted candle inside. I threw a pillow so well that the boy and candle upset, and the boy himself was so frightened that he screamed loudly and brought up the master, and we all got a tremendous lecture.

"I was active and strong, and could throw a cricket ball further than any boy of my age, and was soon very much devoted to the game. I never became scientific, but was a hard hitter and a fast bowler; and my last stroke before giving up the game altogether was breaking a window of the parlour of Lord's cricket ground.*

"I was fairly quick in elementary scholarship, and in mental arithmetic was decidedly beyond the average of boys and men —a gift which was of great convenience and value in after life.

"It was not my good fortune to be at a great public school or at either of the Universities.

"I was always deeply interested in engineering, both in books and works, and was so fixed in my determination about my future career that at the early age of sixteen I persuaded my father to allow me to become a pupil to Mr. J. T. Leather, who was engineer of the Sheffield Waterworks. This was fortunate for my future career, as I had a thorough training

---

* A letter of congratulation from his old friend Mr. Bernard Wake, written to him on his being made a baronet in 1890, recalls to Sir John's recollection "their sprightly movements to Hyde Park, Sheffield, to play cricket at six o'clock in the morning." This "peep-of-day cricket" Mr. Wake dates about the year 1834.

in waterworks engineering, and set out and superintended in
my capacity of pupil the Rivelin and Crookes reservoirs of the
company, and all the business of pipe testing and pipe laying
in every detail.

"During my pupilage I was frequently at Leeds with
Mr. Leather's uncle, who was engineer of the Great Aire and
Calder Navigation, Goole Docks, etc., to give him assistance
when he was much pressed with professional work.

"The result being that my early training was exclusively
waterworks and hydraulic engineering; and before I was
nineteen I was a good engineering surveyor and leveller, could
set out works, and measure them up for certificates to be paid
to contractors."

This record of a healthy, happy boyhood, of a brief
and by no means richly endowed school career, to be
followed by a cycle of busy and yet joyous apprentice
years, is commonplace enough. "Scientific" athleticism
had not yet become a part of the school curriculum,
and juvenile philanthropy was as yet unknown. There
was no sign then of that introspective melancholy and
sentiment which leads boys of this later generation to
expend their energies in the study of social problems.
Young Fowler's surroundings, if not romantic, were sane
and healthy. There were no misgivings in the air as to
the soundness of the economic foundations of society.
The future successful captain of industry was not one to
allow his mind to be "sicklied o'er with the pale cast
of thought," and if misgiving ever occurred to him, his
robust common sense would probably have told him
that the career of a successful engineer (and of that
his confident nature never doubted) would contribute
more to the happiness and well-being of the world
than many projects of philanthropy. To bring pure
water to towns, to design works which would give

employment to thousands, to abolish distance by means of improved locomotion, to make a home that should be a centre of domestic happiness, were objects which at that day appealed to and satisfied the generous instincts of youthful ambition. The air of Yorkshire and the neighbourhood of Sheffield were not favourable to brooding over the *Welt-Schmerz*. When Thoreau went out from the commercial atmosphere of an American city for which he had no taste, and set up his abode in a hut by the Lake of Walden, a life which he has immortalised in a charming fragment of autobiography, there were those of his fellow-townsmen who asked why, if the superfluities of civilisation were distasteful to him, he did not support an orphan. Such questions and such problems were not raised in Yorkshire in the early thirties. Young Fowler became an engineer because in the immediate future that profession seemed likely to be busy beyond all others. Neither sentimentalism nor an abnormally developed athleticism tempted him to decline his part in the workaday world, and the people of Sheffield, at that time at all events, did not put forward the claim of orphans. He became an engineer because it was the obvious thing to do, and he thoroughly enjoyed it.

In addition to the " elementary scholarship," to which allusion is made, there certainly was added a careful and reverent study of the Bible. Like many greater and smaller men, Fowler all his life was reticent on religious subjects, but throughout his correspondence, more especially in the letters written from Egypt many years afterwards, there is abundant evidence that he knew his Bible.

His scholarly equipment may have been narrow,

viewed in the light of modern educational theories; but it was sufficient to enable him to write a terse and vigorous style. The corrections which occur in his familiar correspondence are convincing evidence of a true literary instinct. Lucidity in even his most hurried composition is never wanting, and such corrections as are made are directed to the attainment of greater simplicity in the structure of a sentence and to a severe excision of redundancy of expression. As a result, his all too infrequent contributions on engineering matters to the Press and to periodical literature are admirable examples of popular scientific exposition.

It is a tradition in Sir John Fowler's family that as a child and a boy he was remarkable for his destructive habits. As the principal work of his life was to be construction on the largest and most successful scale, the contradiction is quoted to illustrate one of the generalisations of Froebel, viz. that the destructive faculty of youth can be trained and converted into the constructive talent of the adult and fully civilised man. Carefully considered in the light of evolutionary theory the paradox becomes a truism. Civilisation is the record of the conversion of those human instincts which Professor Huxley has described as those of the tiger and the ape into qualities appropriate to, or at least not incompatible with, our associated life, and it is a fact well known to biologists that the young of species show, even physically, traces of remote ancestry which disappear or become modified past recognition in the fully grown adult.

Fowler's childish success as a story-teller attests his youthful power of imagination. It might seem, and, indeed, he almost suggests it himself, that this talent received no further development. This we believe to be a mistake. Some years ago Professor Tyndall chose as the subject of his presidential address to the British Association the use of the imagination in science. We do not think that a better instance can be found for illustrating the usefulness of the imaginative faculty in science than that which is contained in the history of engineering.

"The civil engineer," says Sir John Fowler in his presidential address to the Institute, "should have a correct appreciation of the objects of each work contemplated as well as their true values, so that sound advice may be given as to the best means of attaining them."

To do this successfully demands some exercise of the imaginative faculty.

The imagination of the novelist and the poet, the "makar," as the old Scots phrase runs, has at its disposal time and space and powers natural and supernatural. The engineer works within narrower limits. Like the artist, he desires to produce a given effect; he has to consider different combinations of material and alternatives of design, and at times he has to contrive some new and previously unattempted expedient, and then (and this is by no means the least important part of his function) he has to clothe his conception in language and form which shall be convincing to capitalists, to railway directors, and, if compulsory powers are needed, to Parliament.

To use the language of a current controversy, it is largely the "ability" of the engineer which sets in motion the vast operations of modern industry. At the same time that ability is limited by the materials available, by the demands of the public, by the willingness or unwillingness of the capitalist community to make the gigantic effort which is necessary for the vast conceptions of modern engineering, and generally by the circumstances of the moment.

The true nature of the part played by the imagination of the great engineer in the history of progress can be well illustrated if we regard the confluence of causes and suggestion that led to the making of the Forth Bridge.

The comparatively simple principle of the cantilever, known and utilised in the oldest form of bridge building, the supply of cheap steel by means of the Bessemer process, the existence of wealthy trading populations on each side of a great tidal estuary, the convenience and economy of conveying goods and passengers over this without " breaking bulk," the existence of great railway companies with vast traffic ready to be carried over the bridge,—in a word, the demand for such a bridge,—the readiness of resourceful contractors and armies of experienced workmen, the existence of a tribunal capable of appreciating the value and practicability of the proposal, and authorised to grant the necessary compulsory powers—such was the " hour." The man or the men who added to this accumulation the vital spark of their ability were the engineers whose business it was to explain how under these conditions the thing was to be done.   In that vast subdivision of labour which is characteristic of modern industry, the part which is assigned first to the imagination, and then to the practical skill of the engineer, is as honourable and as truly originating as is, in the nature of things, vouchsafed to any form of human effort.

Fowler's connection with the Leathers was a very pleasant one, and was the means of introducing him to the great industry of railway extension, then in its infancy. In a letter to his father he speaks gratefully of Mr. J. Towlerton Leather's kindness, and adds that he looks on him as his *pater secundus*. His occasional transference to the service of the uncle, Mr. George Leather, gave him his first experience in railway work, and on an occasion which is of considerable historic interest.

The great George Stephenson, as can now be seen by every tyro, attached at this period what has proved to be an exaggerated importance to ease of gradients, and insisted on taking the line of rail, which afterwards was known as the Midland, down the valleys formed by the tributaries of the Trent, from Derby to Normanton. By this policy the new highway left on one side the important towns of Sheffield, Barnsley, and Wakefield. At this juncture Mr. George Leather was called into consultation by the Sheffield people. The result may be described in the language of the learned historian of Sheffield, Dr. Gatty, the Vicar of Ecclesfield, the editor and continuator of Hunter's *Hallamshire* :—

" Strong opposition was offered to this design, and a rival line proposed similar to that which is now being effected, which would come directly from Chesterfield and Sheffield; whilst the immediate extra cost which would have been incurred would probably have been small compared with that of the expedients to which the town was ultimately driven by the adoption of Stephenson's plan. When the effort to divert the Midland line favourably to Sheffield from its valley course, and the counter proposal under Mr. Leather's advice was rejected, a survey was made of the country betwixt Sheffield and Manchester, passing through Wharncliffe Wood, where much engineering difficulty was encountered. In the severe labour of this survey Mr. John Fowler, then a pupil of Mr. Leather, first acquired his practical knowledge. He assisted Mr. Vignoles in the arduous task of ascertaining the practicability of a route through a district abounding with much natural obstruction, and had ample opportunity of preparing himself for his great local work, subsequently undertaken, of selecting and constructing the large group of railways, now known as the Manchester, Sheffield, and Lincolnshire line."

Fowler's own brief account of the transaction is given in the autobiographical sketch :—

"But a great change in my professional life was to take place before my pupilage was completed. The Stephensons projected the Midland Railway, and as it was proposed to pass at a distance from Sheffield, the Sheffield people opposed it in Parliament, and employed Mr. Leather to represent them, and as a matter of course I was set to work to discover alternative lines to put Sheffield on the main line. But all in vain. Stephenson carried everything before him, and Sheffield was put on a branch. This has since been remedied, and Sheffield is now on the main line of the Midland Railway."

The branch line which first connected Sheffield with the main railway system was the Sheffield and Rotherham line. In a letter addressed to his father in November, 1838, the young engineer, whose professional pride seems to have been wounded, writes with much contempt of the tall talk, as it seemed to him, indulged in by Earl Fitzwilliam and George Stephenson at the opening of this diminutive line.

He has read, he says, with great regret that his uncle had countenanced by his presence the vainglorious proceedings of the opening day. The slight to his native Sheffield, and to Mr. Leather's alternative lines, roused a fine partisan spirit, which must have seemed very diverting to the writer when a perusal of his boyish letter recalled this long-forgotten indignity. Fowler was not a man to attach undue importance to the foibles and errors of great men, of which the history of engineering affords many curious instances ; yet there is a note of pardonable satisfaction and pride in his record of the fact that as a boy he had contended, and not idly or ingloriously, with the great pioneer of railway engineering.

In this connection it is interesting to note that his chief, Mr. George Leather, then so busily engaged in extending facilities of railway transport, had at one time been altogether incredulous as to the value of Stephenson's locomotive.

"Mr. George Leather, c.e.," says Mr. Smiles in his famous biography of Stephenson, "the engineer of the Croydon and Wandsworth Railway, on which he said the waggons went at from $2\frac{1}{2}$ to 3 miles an hour, also gave his evidence against the practicability of Mr. Stephenson's plan. He considered his estimate 'a very wild one.' He had no confidence in locomotive power. The Weardale Railway, of which he was engineer, has given up the use of locomotive engines. He supposed that, when used, they travelled at $3\frac{1}{2}$ to 4 miles an hour, because they were considered to be then more effective than at a higher speed."

Such was the opinion of Mr. Leather in the year 1825, when he gave evidence on the Liverpool and Manchester Railway Bill. *

* See SMILES's *Stephenson*, p. 242.

Fowler's early letters to his father and the autobiographical sketch give us a few incidents and general
descriptions of the work of an engineering pupil in the
thirties which are of considerable interest. The hours
of work were long.

"If I had to go seven miles to the works and back I
walked, as the company could not afford a conveyance. If
I had to get lunch at an inn, the company's limit was one
shilling; and if the work was urgent, to finish drawings or
specifications on a certain day, we constantly worked twelve
hours a day. This was severe early training, but it had much
to do with my success as an engineer in after life, as I was
better prepared to avail myself of opportunities at twenty-
one than most young men now are at nearly thirty.

"But I was strong, and never ill or tired in those young
days, and very full of fun. I remember on one occasion
Mr. James Falshaw (who was one of Mr. George Leather's
assistants, and afterwards Sir James Falshaw, Lord Provost
of Edinburgh) and I were walking in the neighbourhood of
Leeds. I suddenly said to him, 'Falshaw, I will throw you
over that gate.' He replied, 'Fowler, you can't.' I instantly
took him in my arms and did throw him over the gate. He
was fond of telling this story against me when he was Lord
Provost of Edinburgh, and especially to any of my sons."

"At this time," the autobiography relates, "landowners
were violently opposed to surveys being made across their
property, and resisted it by physical force as well as by the
law, and of course we engineers endeavoured to dodge them.
I remember on one occasion I was taking levels along a public
road to accomplish my object, when one of the landowner's
servants stood before my level and obstructed my sight, and
declared he had a legal right to stand where he pleased on
a public road. On this, one of my stalwart assistants inquired
if he also had a right to stand or walk where he pleased on
a public road. Unthinkingly the landowner's man admitted
that any man had such legal right. 'Then,' said my man,

'my right is here, and if you obstruct me I shall remove you'; and walking up to the man, he took him in his arms and deposited him in a ditch. I am not lawyer enough to know which was right in law, but I got my levels. Curious questions of law such as this sometimes arise, which can only be settled by force or demonstration of force."

Force and demonstration of force do not necessarily result in equity, nor in the triumph of the representative of progress. Lord Galway, a redoubtable Nimrod who lived on the borders of Yorkshire and Nottinghamshire, was at this time a violent opponent of the railway system. "Mr. Brundell of Doncaster," says Mr. Grinling in his excellent *History of the Great Northern Railway*, p. 76, "relates that he was horsewhipped by this Lord Galway when surveying for the original London and York line between Blyth and Tickhill."

To return to young Mr. Fowler :—

"On this same survey," he continues, "I arrived in the evening at a well-known inn, and expressed a wish to have dinner. The landlady said she had not a single thing in the house except two geese which were being roasted for one of Mr. Stephenson's assistants, who was engaged on the line, and would soon be back. 'But why two geese?' I said. 'Oh, well,' she said, 'he only cares about one special bit, and it requires two geese to give him his dinner.' 'Well, my good woman,' I said, 'I mean to have one of these geese, and shall take it when cooked; and I am prepared to commit a crime if anyone attempts to prevent me.' She was then ready to give me the goose. I took it, and left a note of apology, which was laughingly acknowledged as being morally, if not legally right."

These incidents, trifling in themselves, mark the long span of time covered by the professional career of Sir

C

John Fowler. When he began life the railway engineer was looked on as an interloper and an outlaw, and, as has well been said in a notice of the subject of this book, recorded in the *Minutes of the Institution of Civil Engineers*—

" An independent professional career, commencing before the railway mania, and extending some years beyond the completion of the Forth Bridge, is indeed a notable record, and it is scarcely possible that one quite like it will ever occur again."

The young engineer appears to have left his father's roof for the first time about the middle of October, 1834, when he was seventeen years of age. The first letter which has come into his biographer's hand is addressed to his father, and is dated from Park Terrace, Leeds, on October 18th, 1834. A few characteristic sentences of it are quoted :—

" DEAR FATHER,—You would be rather surprised in arriving at home on Wednesday to hear I had taken my journeyings to this town; but as I had before told you that it was likely that I should have to come over, you would not think so much about it. The purpose for which I came is the preparing plans, sections, specifications, etc., for a most immense ship lock at Goole—the largest in the world—other locks, bridges, and three miles of canal, all for the Aire and Calder Navigation, at a cost of about £100,000. Mr. George Leather is engineer for that Company, but as yet I have done nothing but copy the specifications—a job I don't much like, but it must be done.

"I have seen very little indeed of Leeds yet, as we work *rather* long hours—seven in the morning to about half-past ten at night—but what I have seen of it, and I believe it to be the best part of the town, has not prepossessed me with it . . . decidedly inferior to Sheffield . . . the shopkeepers not so polite and accommodating as at Sheffield. . . . To-morrow being Sunday, and as I shall not have another opportunity, I intend to go down the railway to Selby, 20 miles. If I get an opportunity I purpose seeing through a cloth factory before I come back. . . ."

The remainder of the letter is devoted to some domestic details as to shirts and the key of his carpet bag.

On November 4th of the same year, 1834, he writes again from Leeds a long letter to his father, who had apparently consulted him about some pumping operations pending at one of Messrs. Greaves' mines in the neighbourhood of his home :—

" I hasten, in accordance with your wish, to answer by return of post. You have not acknowledged the receipt of a letter I sent per Mr. Leather, but I presume it has arrived safe; it was not of any consequence. Mr. Leather as well as yourself know far better than I do of any arguments (though I think it needs none) to show that pumping to the level is preferable to pumping to the surface, but as it is a subject about which I feel considerably interested I feel flattered by your writing. The argument to which I particularly wished to draw your attention, and by you the attention of the Messrs. Greaves, is the amount which might be expended on the level before it would be on a worse footing than pumping to the surface." Then follows an elaborate estimate of two different proposals for draining the mine. " The preceding calculations are not random statements . . . but are the result of calculations (nice and particular ones too)." He concludes with a hope that his father " will pour into Messrs. Greaves' ears such thundering arguments as will not fail to carry conviction to their minds and (oblige them to) see the danger to which they have been exposed."

These two letters record what possibly was our young engineer's first sight of a railway and also the delivery of his first professional report. His employments with Mr. J. T. Leather, of the Sheffield Waterworks, and Mr. George Leather, of the Aire and Calder Navigation, continued without recorded incident till, in June, 1836, we hear of him in Birmingham, where he finds "the population, traffic, and works of all kinds, not merely in the town, but for miles on all sides, most astonishing." The town, he says, is pretty and clean, and there is as much bustle as at Manchester, without any of its disagreeableness.

"We are engaged seeking out a line of railway between Birmingham and Stourbridge, a town about 12 miles from Birmingham, and the last two days Mr. Leather and I have been engaged in taking levels in various directions by way of trying the country; we almost expect the line will have to be continued forwards to Worcester, but we have not yet got particular instructions."

On June 5th, 1836, he writes a somewhat fuller account of his employment as follows :—

"A railway was last year applied for and an Act obtained for a line between Birmingham and Gloucester; but as it passes throughout its whole distance in a barren country (as to minerals), a number of gentlemen in Birmingham are working to get a railway between the same points, but passing through the rich mineral districts of Dudley and Stourbridge and joining the original line at or near Worcester, and have employed Mr. Leather to find out a line for them, and he, conceiving your son might be serviceable, imported him accordingly. But in addition to this competing line, another company has been got up to make a line nearly the same as ours; so that you perceive there is at all events in this case fair competition and no monopoly. We are now engaged taking running levels along the roads, etc., with a view of accurately ascertaining the heights of the different parts of the country, as I should imagine that without exception there is not in England so difficult a country to get a good line of railway through, not even in the neighbourhood of Sheffield."

In these and other employments the young engineer learnt the practical details of his profession, and a year later, a few days after he had completed his nineteenth year, we find him writing from the Birkenhead Hotel on July 18th, 1837 :—

"MY DEAR FATHER,—As I daresay you will have no objection to an account of my present position and progress, I will endeavour to give you a short description thereof. First as to my position. It is at this moment in the coffee-room of the Birkenhead Hotel on the Cheshire side of the River Mersey, and about 1½ miles from Liverpool, having a beautiful view of the river, shipping, and town of Liverpool. Then as to my progress. I will commence with leaving Sheffield on Wednesday and carry you through to the present moment.

"On Wednesday afternoon I left Sheffield per 'Pilot' and called at Sandal, near Wakefield, to see Mr. Dyson (one of the principal resident engineers on the North Midland Railway), armed with testimonials from Mr. Leather, and was very politely received by him, but he thought I had not sufficient experience in masonry to be qualified for the situation I sought, but he should see Mr. Swanwick and Mr. Gooch the following day and would see if he could be of any assistance to me. I then went forward to Leeds by the 'Telegraph,' and in the morning waited on Mr. George Leather, to whom I stated my object and Mr. Dyson's opinion of my competence, on which he desired his son (Mr. J. W. L.) to write to Mr. Dyson and say that I *had* seen a good deal and would, he had no doubt, be able to fulfil the situation creditably. I then returned to Wakefield, but was obliged to wait till night before Mr. Dyson

and Mr. Swanwick returned to dinner, when I presented
Mr. Leather's letter to Mr. Dyson, who however still thought
he could scarcely safely recommend me (on the score of in-
experience), but would mention me to Swanwick and Gooch
and do everything in his power to forward my interests. Thus,
then, as far as the North Midland and Leeds and Manchester
were concerned, I had done all in my power and must wait the
issue. On the following day, therefore (Friday), I went up to
Beeston Park to see George Leather, and spent most of the
day very pleasantly."

The next two days, we are glad to find, he made
holidays, and spent in rusticating with some of the
Leather family at Ilkley, "celebrated for its pure air,
romantic scenery, and cold baths," but on Sunday he
was off again to Manchester, with a letter of introduction to Mr. Locke from Mr. George Leather.

"At Manchester I went to see Mr. Thomas Brittain, and in
the afternoon we had a long ramble in the environs and town,
and at night he kindly offered me a bed, which I, with my
wonted good nature in such cases, accepted, and on Monday
morning, having furnished myself with a map of the town, I
thoroughly examined the public works, particularly the Bolton
Railway, now in process of execution. In the afternoon
Mr. Brittain obtained me the sight of a spinning and weaving
factory, where we saw the whole progress from the raw wool to
finished calico, and in perfection too, being the largest of the
kind in Manchester, which I assure you is saying no little.
"At five o'clock I left Manchester for Liverpool, much
gratified by Mr. and Mrs. Brittain's kind hospitality. The first
26¾ miles on the railway we traversed in sixty minutes, inclu-
sive of five minutes' stoppage, being at the rate of 29 miles in
an hour."

At Liverpool, finding that Mr. Locke was absent,
he determined to await his return, and in the mean-

time went off to amuse himself by an inspection of
the docks, "a rich treat." The letter ends with
"kindest love to mother, yourself, and the innumerable
family at large."

No record is extant of his interview with Mr. Locke,
but we next hear of him writing from Tapton, near
Chesterfield (the home, it may be noticed in passing,
of George Stephenson), where he had secured lodgings
in the house of a Mr. Dean, an old-fashioned farm
"with monstrously low rooms and stone floors; the
bedroom is pretty well, but has a confounded plaster
floor." What it was that took him to Chesterfield
at this date is not stated, but he had now definitely
entered the employment of Mr. J. U. Rastrick, and
in the beginning of 1838 we find him writing to his
father to say that he is duly installed into his new
situation and his new lodgings in the Crescent,
Birmingham.

"Of course," he adds, "I can't say yet how I shall like the
situation, or whether the situation will suit me, but I hope all
will be right. I was very much pleased to meet in the office
with a Mr. Turner, whom you will recollect in breeches and
leggings surveying on the Sheffield and Manchester Railway.
We have an office to ourselves, and he is a first-rate designing
draftsman, highly educated, well informed, and very agreeable, so that I am of course quite pleased with the accidental
meeting."

John Urpeth Rastrick, young Fowler's new employer,
was one of the leading engineers of the day. He was
born in 1780 and died in 1856. He is frequently
spoken of in the railway literature of the period as
Mr. Rastrick of Stourbridge, where he was partner in
the great iron foundry of Bradley, Foster, Rastrick,

and Co. As early as 1814 he had taken out a patent
for an engine, but, like most of the older engineers,
he was at first incredulous as to the merits of the
locomotive engine. In the great parliamentary struggle
for the Liverpool and Manchester Railway in 1825,
he was employed by the promoters, and was the first
witness called in support of the railway. Mr. Rastrick
on this occasion spoke favourably of Stephenson's locomotives then employed on the Killingworth and Hetton
railroads. The Bill on this its first introduction was,
as all the world knows, withdrawn. In the next and,
as it proved, successful application, the precise nature
of the means of traction to be used was left an open
question. Strange as it may appear, when the Bill for
this pioneer line of railway was passed, it was not
yet determined whether the trains were to be drawn
by a locomotive or by a stationary engine. The construction of the railway went on, and Mr. Walker of
Limehouse and Mr. Rastrick of Stourbridge — two
engineers of the highest reputation—were instructed
by the directors to visit the Darlington and Newcastle
Railways, and to examine carefully both plans—the
fixed and the locomotive—and to report the result.
Messrs. Walker and Rastrick reported in favour of the
fixed engine, and "in order to carry the system recommended by them into effect, they proposed to
divide the railroad between Liverpool and Manchester
into nineteen stages of about a mile and a half each,
with twenty-one engines fixed at the different points
to work the trains forward." It is matter of history
how Robert Stephenson and Joseph Locke, under the
direction of George Stephenson, successfully combated
the views of Messrs. Walker and Rastrick, and how

the directors in their dilemma offered a prize of £500 for the best locomotive engine; how this reward was voted by the judges, of whom Mr. Rastrick of Stourbridge was one, to Stephenson's famous "Rocket," and how the triumph of the locomotive was thus finally won.

All this happened some few brief years before young Fowler entered on his apprenticeship. His chiefs, both Mr. Leather and Mr. Rastrick, had in that short period been swept into the full current of the enterprise that had been set flowing by the genius of their old opponent, George Stephenson.

Mr. Rastrick continued to be employed in many of Stephenson's enterprises, and in all probability Fowler's introduction and visit to Mr. Locke led to his being passed on to Mr. Rastrick, then working in close co-operation with Stephenson.

Fuller details of his work are given in a letter to his father, written from No. 1, Crescent, Birmingham, on February 22nd, 1838 :—

"I have now been more than a fortnight in Birmingham, but we have been so busily engaged that the time has appeared much shorter. However, I think I have seen sufficient to be able to say that everything is proceeding satisfactorily and will prove much to my advantage.

"I told you, I believe, in the short, hurried note you would receive when my uncle returned to Sheffield, that we were preparing working drawings of the works on the Manchester and Birmingham Railway, and London and Brighton. Since that time we have lost the Manchester and Birmingham by a stock-jobbing trick, at which the Manchester speculative swindlers are such adepts. On Saturday last we were making great exertions to complete some designs that Mr. Rastrick was to take with him to Manchester on Monday morning for a meeting of directors. On his arrival there he was told that they (the directors) had given notice for an alteration of the line of railway from beginning to end. Now Mr. Rastrick had never heard one word of their intentions, although he had come from London a few days before with the very person who had been there to make arrangements. Not one word was said to him on the subject. He was naturally very indignant, and told them (the directors) that as they had chosen to take such a step without consulting him he should have nothing more to do with the concern, and from that moment their connection must cease. This, as you may naturally conceive, rather astonished us, for Mr. Rastrick in his usual blunt way told us on his return, ' Well, it's no use drawing any more bridges on the Manchester and Birmingham; we must now attend to the London and Brighton.' What will be the upshot of it all I can't tell, but I shall not be at all surprised if we have to oppose the application for their new intended line.

"We are therefore attending to the London and Brighton Railway, that is, drawing working plans of the bridges, etc., required on the line."

His stay in Birmingham was not, however, to be long. On March 3rd, *i.e.* in less than a month from his arrival, Mr. Rastrick sent him to London, and he writes on that date to his father, saying that he has enjoyed an inside seat in the " Emerald " Coach.

"Although my residence will be changed from Birmingham to London, I am not aware that that circumstance will make any material alteration to me personally with respect to my engagement with Mr. Rastrick, but it may be of advantage in other points of view; and as it is very probable I shall stay there during my engagement at least, I shall have an opportunity of seeing London, and maybe obtain a wrinkle from southern acuteness. As far as personal comfort is concerned I am afraid the change will be for the worse; I don't think

I shall get such comfortable lodgings as Mrs. Farrow's, nor shall I have such very commodious offices, and of course I don't know a single individual in town, but as personal comfort is not any part of the principle of Civil Engineering these things are of no importance."

His friend Mr. Turner, he is glad to say, is going with him, and they hope to continue together their occupation of designing the bridges for the London and Brighton Railway.

The following letter gives an account of our young engineer's first journey to London, at that time considerably behind the North in respect of railway development :—

"CRAVEN HOTEL, CRAVEN STREET, CHARING CROSS,
"Monday, 6 p.m. (*March* 5th, 1838.)

"DEAR FATHER,—I arrived at last in that wonderful centre of everything that is great, good, bad, and indifferent, that receptacle of the most accomplished swindlers and the most enterprising men of business, London, after a most tedious and protracted voyage of *eighteen hours*, the roads so positively bad that, although we had six horses nearly the whole of the journey, the passengers were obliged to dismount and the coach dragged at a foot's pace for almost a mile at a time, and at no very distant intervals ; however, here I am at last, and I must beg leave to say that since my arrival I have made pretty good use of my time. I have seen St. Paul's, Westminster Bridge, Blackfriars and new London Bridges, Post Office, etc., etc., been swindled by a cabman, called on 493 different houses for lodgings, and at last engaged a most beautiful sitting-room and bedroom in a very excellent situation, Warwick Court, Holborn. The house is occupied by a respectable architect and builder, Mr. Smith, who appears quite a literary character, and has a good library, a free use of which he kindly offered me. I have engaged the rooms for a week certain, at a moderate, a very moderate rent, considering the situation and style, for the sitting-room is positively elegantly furnished. I intend to keep my own tea and coffee, etc., but not to dine at home, but at a chop and eating house, where to-day for 1s. 2d. I had one of the best dinners I ever ate.

"I have no time to say much in this letter, for the post will soon be closed; however, I must not forget the most important subject which I have to mention, and unless for it I should not have written so early as to-night. The fact is, I have only about 15s. left, and I don't like to ask Rastrick so very soon to advance on account; it looks rather too snobbish. However, I hope you will have received some portion of the money due to me from Mr. Burbeary. I must beg positively of you to send a £10 Bank of England note enclosed on Tuesday and directed to me at care of J. U. Rastrick, Esq., C.E., 454, Charing Cross. I must again beg you will not neglect, because if a man can't meet his liabilities he becomes a bankrupt. . . . I think I am settled here for the next five months at least, which is just what I wanted, but I was rather disappointed to-day when I tried Rastrick to increase my salary. He told me he had had so many persons applying for employment on the Brighton Line (the crack of all other lines) that he had been offered to be served for *nothing*. But I don't care much; salary is not my object now, and what I have will pay my expenses pretty well, with economy.

"I have a great deal to say to you and my mother, and if I stay here you may expect, on the payment of 10d. postage, to receive some descriptive accounts of London and its wonders.

"Give my kindest love to my mother, and tell her that I thought of her when I first set eyes on St. Paul's Cathedral."

In a long letter written to his grandfather he gives the following account of the nature of his work in Mr. Rastrick's office in Birmingham and then in London :—

"I must now, in as short and succinct a manner as possible, tell you how I spent my time, which I assure you was anything but in idleness. In the morning at nine o'clock the office opened, at half-past two we dined, returned at half-past three, and remained until seven. These were our office hours, and as to my employment in the office I am afraid you will not understand that part of the business quite so clearly; however, to give you an idea, Mr. Rastrick (the gentleman with whom I am at present engaged) is the principal engineer for a line of railway from London to Brighton which is now in progress of being executed, and, as you will imagine, a great deal of work has to be done in making bridges over and under roads, over rivers and canals, etc., and as these works have all to be let by contract, and consequently very particular working drawings made of all the works required, you may conceive that a great deal of preparation is necessary before this can be done, and the work commenced. Now my part or employment was to make designs for bridges, calculations, etc., which is, of course, very important, as immense sums of money might be thrown away without due care and consideration.

"Mr. Rastrick, finding it very inconvenient to be at such a great distance from the line of railway on which we were engaged (the London and Brighton), very properly decided to remove us nearer to our work, especially such of us as were engaged in matters requiring constant reference and communication with surveyors, etc., employed taking surveys and levels along the line. This was the reason I was removed to London. . . . I occupy two rooms, the sitting-room, which is on the ground floor, is a very well furnished room, with couch, looking-glass, stuffed arm chair, paintings, etc., and the lodging-room is a comfortable room, though near the top of the house. I pay a certain sum per week for the rooms, and then have to find my own coals, candles, tea, coffee, sugar, milk, wood for lighting, washing, butter, bread, etc.

"Our office hours in London are from half-past nine to half-past five, with the intervention of a dinner-hour, which leaves us a tolerably long evening to our own disposal; and consequently, after leaving the office I go to a coffee-house, and for 1s. or 1s. 2d. I get as good a dinner as I could possibly have, and see the daily papers into the bargain, after which I generally look a little about some part of London and then return home, where I amuse myself with either studying, reading, or writing, according as I may feel myself disposed at the time, and on the whole am as comfortable as can be expected considering I have no society and have to pay 4s. 9d. per ounce for coals.

"I have not yet seen much of London, excepting the bridges over the Thames, St. Paul's Cathedral, Westminster Abbey, Regent's Park, Day and Martin's blacking manufactory, and a splendid gin palace; the two last places, notwithstanding their odd titles, are decidedly more magnificent places than the new post office in Sheffield."

On May 11th, 1838, he writes from 45, Stamford Street, Blackfriars, to his father with regard to some local road-making plans on which our young engineer appears to have been consulted. The letter is curious as illustrating the warm and personal way in which he espoused his client's cause, and also as recording the first occasion on which he appeared as witness before a Parliamentary Committee.

In the course of the letter he comments very freely on what he considered the shady tactics of the opposition, and on some mismanagement on the part of his friends. He approached Mr. Rastrick with some trepidation, "he is such a queer fellow," and asked leave to attend the committee. Rather to his surprise, for the office was very busy, Mr. Rastrick told him "to take and go," "and accordingly I did take and go." Under date June 18th, 1838, he thus describes his first appearance in the witness-box :—

"I had a long cross-examination to-day, and succeeded much better than I anticipated, for I was not the least nervous or confused, and consequently passed very respectably, I think, but I must say I think our counsel (Mr. Talbot) did not make the most of his case.

"Baines was against us, but he was rather lenient with me than otherwise, and did nothing more than his duty to his clients.

"I have now made my maiden bow, and know the worst of this much-dreaded cross-examination, but I have been obliged to give up a journey to Manchester respecting the Oldham Waterworks, which is now in the House, and is in an awkward situation from the surveyor making a complete failure in the Commons and losing the cause there. Mr. G. Leather, of Leeds, is the engineer employed, and begged permission of Mr. Rastrick for me to go down and repair this blundering, but this road affair compelled me to stop in town."

It is evident that young Mr. Fowler's chiefs were beginning to recognise his latent capacity.

In the same letter, viz. May 11th, 1838, Fowler gives an interesting account of a visit paid to Maidenhead Bridge, Brunel's beautiful construction, about which grave misgivings were then rife :—

"On Sunday morning last, at eight o'clock precisely, I threw myself on the outside of an Oxford coach to go to Maidenhead to see a brick bridge built there over the Thames, on the Great Western Railway. The day was beautiful, and I very much enjoyed the journey. Maidenhead is 26 miles from London. . . . The bridge I went so far to see is of a novel construction, being built of brickwork set in cement for two elliptical arches of 128 feet span each, and is now, I am sorry to say, in a dangerous situation. The centering has been slackened, and the arches have followed it for 5 inches at the crown, and now rest on them, but not as you would suppose at the crown of the arch, but at about 15 feet on each side, while the crown is 2½ inches clear of the centering. This settling has made the outline of the arch perfectly Gothic and totally destroyed its symmetry, but the worst of the matter is that the arch which has been turned in half-brick rings has separated at every other joint, and consequently the arch is further there weakened by the lower part of it leaving the upper part. The outside of the arch shows eight half-brick courses, but the inside has thirteen. The spandril wall of one arch has cracked from top to bottom, so likewise have all the walls for backing up the inside of the arch, both occasioned, of course, by the settling. The foundations are on rock, and therefore very unyielding. The centering is a very excellent and a very scientific one, the whole of the timbers being in a state of thrust."

" What will or can be done is too much for me to say."

The incident here narrated is thus noticed by Mr. Brunel's son and biographer.

" The great bridge over the Thames at Maidenhead contains two of the flattest and probably the largest arches that have ever been constructed in brickwork. . . . The main arches are semi-elliptical, each of 128 feet span and 24 feet 3 inches rise. . . . The radius of curvature at the crown of the large arches is 165 feet, and the horizontal thrust on the brickwork at that point is about 10 tons per square foot. . . . The Maidenhead Bridge is remarkable, not only for the boldness and ingenuity of its design, but also for the gracefulness of its appearance. If Mr. Brunel had erected this bridge at a later period, he would probably have employed timber or iron ; but it cannot be a matter of regret that this part of the Thames, although subjected to the dreaded invasion of a railway, has been crossed by a structure which enhances the beauty of the scenery."

With regard to the defective condition in which young Mr. Fowler found the bridge, Mr. Isambard Brunel gives the following information :—

" During the construction of the bridge a part of the crown of the eastern arch proved defective in consequence of the cement in the middle of the brickwork not having set sufficiently at the time when the centering was eased. Apprehensions which had been entertained by some as to the safety of the structure were groundless, for when the defective part was taken out and replaced no further trouble was experienced. The bridge has stood well, and has shown none of those symptoms which an overstrained structure exhibits."

Mr. Fowler's connection with Brunel's beautiful bridge was destined to be renewed again many years later. In 1893, addressing the Merchant Venturer's School at Bristol, he gave the following account of the widening of the bridge. He cites it as an instance of the improvement in the manufacture of materials, other than steel and iron, used in bridges and similar structures :—

" This bridge," he said, " consisting of two main arches, each 128 feet span with a rise of 23 feet 3 inches, was built in 1837 with bricks known as 'London Stocks' and with mortar made partly with chalk-lime, and partly but chiefly with Roman cement and sand.

" With such materials the bridge had very little margin of stability ; indeed, with the east arch considerable difficulty was experienced in putting it into a perfectly safe condition.

" When the widening of the Great Western Railway from London to Didcot was decided upon, the mode of dealing with the Maidenhead Bridge necessarily required special consideration.

" Very naturally, as the consulting engineer of the company since the death of Mr. Brunel, the directors referred the matter to me for my decision, and I had no difficulty in advising the Board to make the addition on each side, and preserve the old beautiful lines and elevation.

" The work is now finished, and is satisfactory to the directors and Mr. Brunel's old friends.

" The bricks used in the widening were of very superior

D

quality, and possessed a strength of resistance against crushing
several times greater than the bricks originally used in the
work, although no doubt Mr. Brunel adopted the best material
at that time obtainable.

"The mortar used for the widening (Portland cement and
clean Thames sand) possessed a degree of cohesive strength
and a regularity of quality, which were unknown in the days
of lime and Roman cement.

"With such superior materials it is not surprising that the
additional arching was a work of perfect simplicity, and free
from the anxiety which the original work under other conditions caused the engineer.

"When we consider the amazingly improved methods of
manufacture of the three common but most important
materials employed upon structures for land, and with ships
for water, viz. steel, bricks, and cement, we are unable to
form any definite idea of the total amount of benefit—
economical and otherwise—that has resulted from these improved methods."

It is an interesting episode, though as introduced
here it somewhat anticipates our narrative. It supplies
a comment on the pessimism which, from an artistic
point of view, is ever bewailing the deterioration of
modern methods of manufacture and workmanship.
The artistic proprieties perhaps are and have been less
considered than they should be, but the superiority of
every class of material at the present time gives the
engineer and the architect a larger scope for the accomplishment of his design. There always is and must be
a question of economy in the background, for the desire
to economise effort is a condition of our existence, but
it is very questionable if it has more weight as against
artistic considerations with the present generation than
with generations which have gone before.

CHAPTER II.

SUBORDINATE EMPLOYMENTS

HIS engagement with Mr. Rastrick was now drawing to a close, and he had to consider the next step. In pursuance of the plan of allowing Mr. Fowler as far as possible to tell his own story, the following letter to his father is inserted.

" 45, STAMFORD STREET,

" BLACKFRIARS ROAD, LONDON,

" *July* 18*th*, 1838.

" MY DEAR FATHER,—My engagement with Mr. Rastrick being now within a few weeks of its termination, I thought it best to have his wishes with respect to its continuance, and to-day I have had a conversation with him on the subject, the result of which I am anxious to communicate to you, and receive your answer per return, if possible, as I have promised to give a final reply this week.

" By way of introduction, I must beg of you not to mention to *any one* the salary I have had or am to have, for I detest the country fashion most heartily of interfering with every person's affairs and business.

" Mr. Rastrick's offer (which I feel disposed to accept) is £160 a year for the next twelve months, or £80 for six months and the chance of a resident engineer, should he be enabled to give me one, and in case of being from home 10s. 6d. a day for ' wittels.'

" Now although this is not a very high salary, yet I think

on the whole it will be better to accept it at the present, until something better turns up, for the following reasons :—

"Baines will not then be able to badger me and express his indignation to the House of Lords at the heinous crime of my being twenty-one years of age.

"If I should be disposed to try my luck in Yorkshire the people will almost have forgotten me in that time, and consequently I shall be much better thought of.

"And, lastly, it is possible I may meet with something here worth having in the meantime.

"Now I am rather disposed to think that Mr. Rastrick would give me rather more than £160 a year before he would allow me to leave, but I would much prefer concluding a bargain with him, with an obligation on his part to give me a better situation if he has the opportunity, than drive the bargain to the utmost sixpence, because I believe Rastrick will do me a service if he can; however, do let me have your opinions on these points.

"I saw my uncle in town last week for a very short time, but I shall be very sorry if he ever comes up without my seeing him; and on my mentioning to him what would probably be the result of my trying again to agree with Rastrick he fully concurred with me in opinion as to the policy of continuing, and you are aware that I entertain a very high opinion of his judgment in mundane affairs.

"I enclose my small account for the Halifax and Sheffield Road Trustees, which I will trouble you to deliver for me with my best compliments to Mr. Burbeary."

Then in his character of elder brother he requests his brother Henry to send him specimens of his work, and with an eye no doubt to his own information as well as Henry's education he offers to pay for as many designs of bridges "as he can afford to do for a sovereign."

A month later, August 14th, 1838, we find a further

letter to his father on the subject of Henry's education,
and it is amusing to notice how completely this young
man of twenty-one has taken in charge the whole of
his family. Henry is about to engage with his own
old instructor, Mr. Leather.

"I hope," he says to his father, "you will not procrastinate
again for the very moderate period of twelve months before
taking the final step, for it is quite possible Mr. Leather may
have the offer of a large premium for so desirable a situation."

He feels unable to advise about Charles, who was
desirous of becoming an architect. He recommends
London in preference to Sheffield, and prays the family
not to be bigoted in favour of Flockton, a local architect
of Sheffield.

The letters ramble on in a good-natured way, confident, yet always respectful and affectionate, and
occasionally jocose.

"The possibility you mentioned in your last of yourself and
mother coming to London was so extremely ludicrous that
I laughed for twenty minutes consecutively. My mother in
London, ha! ha! But I hope she will go with you to the north,
and *if* you should come to town, I should of course be most
glad to see you. . . .

"I want now to say a few words to you about treating me to
a theodolite, a very beautiful instrument which I can purchase
cheap, £16; for with the expense of London and occasionally
purchasing a few books and instruments I am as poor as a New
Poor Law cat."

The future Conservative candidate was apparently
even at this time a Tory in politics, but there are very
few political allusions in his letters of this date. In
this letter he expresses a fear that "our radical

ministers will take some step, perhaps open the ports
to lower the price of corn in a short time, for people
are becoming very clamorous at the rise of bread."

Then at the end of a three-page, closely written
letter the important information is added, "I have
agreed with Mr. Rastrick again on his own terms."

As to his future, he continues to speculate. He
would like, when leaving Mr. Rastrick, to have employment on the Manchester and Sheffield, perhaps as
a contractor, or perhaps Lord Wharncliffe would give
him the post of a resident engineer. His uncle at this
time purchased Chapel Town Iron Works, and was
looking out for a moneyed partner.

"I tell you candidly," he says in a letter to his father, "that
if I could by any means obtain the use of £3,000, I would
negotiate with him to join as an active partner, and reside upon
the works, for although my professional prospects are perhaps
as bright as most young men of my age, yet I am not so confident about the future, and if I had a chance of entering
in the Chapel Town concern I assure you I would embrace
it, but if you don't know any Mr. George Greaves, or some
such person who has a few thousand he is anxious to employ,
I am afraid I shall have to give it up."

By this time he had moved to 25, Frederick Place,
Hampstead Road, but in November he was sent by
Mr. Rastrick to Cumberland, and he writes to his
father from Preston, November 5th, 1838 :—

"As I have half an hour to spare before the coach starts
for Lancaster, I avail myself of the opportunity of writing a
few lines to explain the object of my being here.

"I start from Lancaster to find the best line of railway
through Cumberland near to the sea coast as far as the Calder

River (a distance of 40 miles), where I shall be met by a gentleman whom Mr. Rastrick has sent to start from Maryport.

"This railway will form part of a continuous communication from London through Birmingham, Preston, Lancaster, Maryport, Carlisle, and thence to Glasgow, of which that portion from London to Preston is now open and from Preston to Lancaster is in a forward state of execution.

"The peculiar feature in the railway in which I am engaged here is the embanking of Morecambe Bay from the sea, by carrying the railway over the estuary of Morecambe and reclaiming about 50,000 acres of land from the sea.

"As I expect to have nearly fifty miles of levels to take, I fully anticipate being in Cumberland for a month or six weeks at the least, especially as the weather appears likely to prove unfavourable.

"I am accompanied by a gentleman (Mr. Bristow) who has only been a pupil of Mr. Rastrick for a very short time, and whom Mr. Rastrick has placed under my *judicious* care in a professional view, but he is a very clever, gentlemanly fellow, about nineteen years of age, and a first-rate mathematician, and besides a very pleasant companion.

"The remainder of the line will be under the superintendence of two gentlemen who have frequently taken levels for Mr. Rastrick, and knowing his ways and having two levels instead of my one, will have an advantage so far as point of rapidity."

Then follows a request that his old staff holder, Marsh, should be sent to him.

In his next letter, dated November 12th, 1838, he gives a further account of his undertaking.

"You wish to know a little more of the scheme on which I am engaged, but as I am said by an eminent phrenologist to be deficient in the bump of description, I am afraid I can't be very clear and explicit.

"You are probably aware that a good deal of interest has been excited by the discussions, surveys, etc., which have taken place to prove whether the best line of railway to Scotland should be taken on the western or eastern coast. Now it is of great importance to the existing railways in the direction of the north from London that theirs should be made use of as part of the grand Northern Trunk, and it is with reference to the scheme of continuing the railway communication from Lancaster across Morecambe Bay and along the western coast to Carlisle and thence to Glasgow and Edinburgh that we are now employed. The portion of the scheme which is now being surveyed by Mr. Rastrick is from Lancaster to Maryport, and this distance is divided into two portions: in one from Lancaster to the Calder River beyond Ravenglass, I have the honour of being entrusted with the exploring of the country and finding and levelling the line, and thence to Maryport is completed by two gentlemen who have before levelled a good deal for Mr. Rastrick.

"The principal feature in this scheme, as I think I told you before, is the formation of a railway across the bay of Morecambe, and thereby reclaiming about 40,000 acres of land from the sea. The bay I have crossed to-day with a chain at low water, and then the sands are quite dry for miles in extent and most beautifully level, and indeed at the point of crossing them with the railway embankment (10½ miles in length) the sands are only covered to a small extent at low water.

"I have been detained at Lancaster considerably longer than I expected, partly from the unfavourable state of the weather, and partly from having taken a line quite distinct to the one which has been thought of, and therefore having to level twice as much as was expected; however, I have this morning sent off an immense despatch to Mr. Rastrick with the result of our operations, which I hope will be satisfactory. . . . The whole of the Cumberland hills are covered with snow."

A fortnight later he writes again :—

"MY DEAR FATHER,—I must apologise for not complying with your wishes as to a letter per week, but really I have been too much engaged until this afternoon, when a tremendous gale of wind has given me a little leisure.

"My survey is now almost completed, and if the weather should be favourable I shall be in town again in a week.

"After finishing on the Lancaster side of Morecambe Bay, I had a difficult country to encounter in the peninsula of Furness, and my progress with respect to distance was slow; but I determined thoroughly to examine and find out the best possible line in a general point of view without entering unnecessarily into detail, and, I think, succeeded tolerably well. And at the northern extremity of the peninsula I had to level across the estuary of the Duddon, which at low water is confined to a channel of 150 yards in breadth, but at high water is nearly two miles. Now although levelling across sands is generally a very simple operation, yet in this case it was not so particularly simple.

"In the first place the sands are almost what might be called quicksands, and in the next place low water was before daylight and we were two hours after the ebb, and in the last place I had an attack of slight English cholera or some other confounded affection.

"However, we managed pretty well over (being ferried across the channel in a boat) to the opposite shores, but the tide was then rolling in apace, and we had nearly a mile to run to get to the boat (now pray don't be alarmed, I beg, for I assure you I am not going to be drowned). However, off we started at a rattling pace; the first low place which had been covered by the tide was not more than a foot deep, and easily passed; then, tally ho! on for the next, which we knew was the deepest, and if we couldn't pass it return was impossible; however, the first got over in about 18 inches of water, the next, who was at his heels, about 2 feet, the next deeper; I was the next with my level on my shoulder, and got clear of a tremendous wave, which, however, caught

Marsh, who was at my heels, and nearly overset him. And thus all got clear with merely a desperate cold bath, then off and away over the dry, intervening high sand to the boat in the principal channel, which by this time had become more than half a mile broad; and being landed on the opposite shore we made the best of our way to quarters and changed, and fortunately I took no increased cold.

"After sending off despatches of this work to Mr. Rastrick, I passed over into Cumberland, and, the country being favourable, I have progressed at a rapid rate until to-day, when the wind has been too high for levelling.

"Yesterday, after attending divine service at Bootle Church, I and Bristow ascended one of the mountains (Black Combe), 2,000 feet high, and the day being very clear we had a most magnificent prospect, the inland Cumberland mountains (Helvellyn, Scawfell, Skiddaw, etc.) being snow-capped in a most splendid manner, and besides other objects we saw the Isle of Man and Scotland very distinctly.

"I wrote yesterday to Mr. Rastrick for some more ammunition, as we find we shall be obliged to remain in the country if not taken out of pawn. Our expenses have been enormous, being obliged to post very long distances round Morecambe Bay, Duddon, etc., but we have been *very careful* as far as possible. However, £50 in a month—or rather a little more than three weeks—looks almost as bad as valuators travelling with four horses on the Manchester Railway, seeing double hills, and estimating the wrong ones at a distance that required very powerful optical instruments even to distinguish.

"As the supporters of the West Cumberland and Furness Railway are very sanguine, it appears probable that they may 'deposit' in March; and if so, after assisting Mr. Rastrick in making his calculations and report, I may perhaps be down here again in a month to take the parliamentary levels."

His next letter is from Ravenglass, Cumberland, under date December 4th, 1838 :—

"I have nothing particular to relate except my having

been most completely soaked through and through every day since last Sunday but once, and getting amongst the water on the tide rivers with which the shore here abounds; but I am now quite accustomed to it.

"We have been so much delayed beyond the time at first contemplated for our getting finished, that I have been devising schemes for working all weathers in future, if possible, for Cumberland wind and rain set all calculation at defiance.

"I had a long interview yesterday with Sir Fleming Senhouse, who takes the principal management in this scheme, and lunched with him. I was not aware before that I was Mr. Fowler, *the* engineer; but so Sir Fleming said, and I allowed it to pass."

Mr. Fowler at this period of his life looked considerably more than his real age. The crime of being only one-and-twenty was not very patent, and in after years Mr. Fowler used to say that the fact was to his advantage.

Then for his father's information there follow some remarks on the state of agriculture, the nature of the fences, which here consisted of soil embankments of about 5 feet high, 4 feet wide at the top, crowned by a low hedge, and with ditches on both sides "in size more like young canals."

Three days later, December 7th, 1838, he is on his way back to London, and writes to his father from Preston :—

"I expect, if the West Cumberland and Furness Railway scheme should go on, we shall be very busily engaged from the present time to March 1st (the last day of depositing plans for Parliament), but I have been so long from headquarters that I am rather ignorant of the arrangements which have been made on the subject; but I will take an early opportunity to write you a full account, but I quite expect I shall not have the pleasure of eating my Christmas pie at Wadsley Hall."

On the 19th December he writes again to his father that his survey has been approved by Mr. Rastrick, who was then considering his report.

"My survey was very satisfactory, and all the deviations I took from the general principle of the line Mr. Rastrick marked out upon the ground (I don't mean questions of detail) he has adopted without exception, and if the scheme proceeds I expect to go down again immediately and remain till March. I don't think I shall be able to get down to see you this Christmas, but I don't exactly know, as there will necessarily be a calm for a week or two during the time Mr. Rastrick's report is being printed, a meeting called, and the question of further progress decided on before the *hurricane* commences, after which casting anchor is out of the question, for, once embarked, we must ride out the gale. If I do come, I shall dine with you on Sunday next, but don't expect me until you see me."

Whether Mr. Fowler spent Christmas with his father or not we are unable to say, but he returned again before long to his surveying in Cumberland. The following letter, written in 1883, relates an incident in connection therewith :—

"I remember," he says, "the incident in crossing Morecambe Bay sands perfectly well; indeed, it was a danger not likely to be forgotten.

"Mr. Bischoff and I, and Mr. Jonathan Binns, of Lancaster, were driving across the sands when I was acting as Mr. Rastrick's chief engineering assistant in laying out the railway from Lancaster across the sands thence near to Barrow (now very famous for iron works, shipbuilding docks, etc.), through

Furness, across Duddon sands, Ravenglass, Whitehaven, and Workington.

"The railway was called, I believe, the Morecambe Bay Railway, and the plans were deposited in 1838 and 1839. Mr. Bischoff, Binns, and myself were posting across the sands when a sudden and violent snowstorm came on.

"I recollect that, although the windows were closed, the fine snow, driven by a strong wind, penetrated into the carriage and formed a small snowdrift inside. In those days the route across the sands, and especially the tidal channels, which had to be crossed, was marked by boughs of trees put into the sand as guides to travellers, and men were engaged to watch any changes in the channels, and then alter the position of the boughs. To show that this was necessary, it was narrated that a coach got into a channel not many years before, and all were drowned.

"After passing some distance across the sands on our journey, we became aware that our horses were no longer trotting, but were walking slowly; and on opening the window and speaking to the driver, I found he was crying, and admitted he had lost his way.

"As the youngest of the party, I got upon the box, took the reins, and endeavoured to get an idea of our course. I had to take a leap in the dark, and changed the course considerably, but as it turned out not sufficiently, for a man soon after galloped up to us and inquired where we were going. Of course we answered that was exactly what we wanted to know.

"The man was a carrier, and had detected the spot where our driver had gone wrong, and where I had partially made the mistake. He soon brought us to shore.

"It appeared that the driver had turned his course, so that he was driving out to sea, and I had corrected it so that we were driving parallel to the shore. In one hour we should all have been drowned except for the carrier.

"Binns is dead long ago, but Bischoff lived to resuscitate the Tilbury Railway which I made many years ago, and I lived to make the Metropolitan and Metropolitan District Railways, and a good many other things.

"The Morecambe Bay Railway was an important and thoroughly sound conception, but it was proposed before its time . . . I was very proud to have been entrusted, at twenty-two years of age, by Mr. Rastrick with the sole responsibility of laying out so great and difficult a work!"

The West Cumberland and Furness Railway, or the Morecambe Bay Railway, as Sir John Fowler calls it, apparently came to nothing, but in June, 1839, Mr. Rastrick sent his competent young assistant to Burton-on-Trent to take the levels of a line of railway, a continuation of the Manchester and Derby, to join the Birmingham and Derby at Willington, some five or six miles from Burton, but, as he wrote to his father :—

"It is merely an opposition scheme to the Manchester Extension Railway now before Parliament, and not a *bonâ fide* undertaking."

On June 29th, 1839, he writes to his father :—

'I shall most certainly leave Rastrick in August, when my engagement expires, but shall not decide exactly what to do until I come down into Yorkshire, which I have considered will be best in several respects, although probably I shall immediately open offices in Sheffield as 'Civil Engineer and Surveyor,' but I have several reasons for wishing it not to be spoken about at present. Your assistance, which would be indispensable, I take of course for granted from your usual kindness, would not be wanting.

"Before I leave London I purpose getting a few instruments and books of reference, which I can do much better and cheaper on the spot than if I have to send for them, and for which I should want your assistance, and my time being now rather limited, I wish you would write as soon as possible.

"The amount about as under :—

	£	s.
5 in. Theodolite . . .	26	5
Pocket sextant . . .	5	5
Pentagraph	9	9
Protractor	2	2
Books, including Smeaton's reports, Tredgold's works, etc. that I have marked down in a catalogue . .	18	6
	£61	7

"Even if I should not open offices in Sheffield, the above list will be necessary in almost any situation I should be in. . .

"We are much engaged in Parliament just now in opposing the Manchester and Birmingham Extension Railway Bill, and from the present state of the promoters and opposers it appears likely to be one of the most expensive applications and oppositions ever made, not less I should say than £200,000, which would make a snug little railway of itself."

A letter of August 3rd, 1839, gratefully acknowledges the receipt of £70 sent to him for the modest equipment above mentioned.

"My time," the letter goes on, "with Rastrick* is now very short, as I shall leave him next Thursday, and, the day after, I propose going down to Potter (James Potter, Esq., Civil Engineer, Borden's Farm, Balcombe, Sussex), and after remaining with him a week or ten days, return to London, and then after settling my important worldly affairs, you may expect me down in Yorkshire. The astronomical calculations of the time of appearance of this extraordinary comet give August 24th as the time when he may be expected to be seen in this hemisphere."

* In a letter to the *Times*, September 26, 1891, Sir John Fowler drew attention to the omission of Mr. Rastrick's name from an article commemorating the jubilee of the London, Brighton, and South Coast Railway. He seems throughout life to have cherished a kindly recollection of his old employer.

Eight days later his plans had again changed, as we learn from a letter of August 11th, written from Leeds :

"On Wednesday last I received a letter from Mr. J. T. Leather, stating that his uncle was in want of someone to be on a railway of his in Durham, and wishing me either to come or write immediately on the subject.

"I consequently left London on Thursday night, as I conceived writing would be a very imperfect mode of obtaining proper and correct information of the nature of the situation, salary, etc., and it appeared it was necessary to be decided in the shortest possible time.

"On my arrival in Leeds, and calling on Mr. Leather, I found the situation he wished me to take as follows :—

"To be at Seaton, in Durham, on the railway, and his office at Leeds alternately, and the salary he could give £200 a year.

"Now as this is something different to what you expected me to do, and, in fact, what I intended to do myself, I will detail to you the circumstances which influenced me in deciding on the step I have taken, which is to accept the appointment.

"Mr. Leather's office is a most excellent one (from the varied and miscellaneous work presented) for improvement, and some of a character with which in a great measure I am unacquainted, and the railway in Durham has sea embanking and other peculiar works which make it desirable, and I am desirous of getting yet more information on works. And generally there are several things which will be better obtained in his office than anywhere else, before I commence business on my own account; and in future I am convinced the only way for a man to be successful is to understand all the scientific and practical part of engineering *well*.

"I have not decided on this step either from indecision of character or in haste, but from a well-considered balance of advantages and disadvantages.

"I certainly only propose to remain a year with Mr. Leather, and intend to open offices in Sheffield on the 1st September, 1840, but this is not what I should like Mr. Leather to know, as he might not perhaps be pleased."

In the autumn of this year we find him settled down in permanent quarters at Mr. Stonehouse's farm, Greatham, near Stockton, Durham.

"Mr. Stonehouse," he writes, "is a farmer of 400 acres of land, and I have exceedingly nice rooms, and they appear a very good sort of people; but of course it was rather a favour to get here as they are not in the habit of taking lodgers. However, that was done through the kind offices of one of the directors.

"I find myself very fully occupied, which must be a sort of permanent excuse if I write a short letter, or if I don't write at all, for just at this moment I have made an engagement with the contractor to be at one end of the work (five miles from here) at seven in the morning, to measure up for his monthly payment, and at twelve I have to meet some of the directors, the solicitor, and the agent of the Dean and Chapter of Durham about the alteration of some roads; afterwards I have to put some levels in for a part of the work."

Mr. Fowler, during the years 1839 and 1840, remained in superintendence of the making of the Stockton and Hartlepool Railway, spending his time between Greatham and Hartlepool, with occasional absence at Leeds, the head office of his employer, Mr. Leather.

The following letter to his principal, a draft of which has been preserved, illustrates the sort of work which had to be performed by a resident engineer.

"GREATHAM, *July* 30*th*, 1840.

"We have had a considerable influx of bricklayers during this week, and most of them are first-rate hands from Leeds, so that now we have men for the outside work competent to make the most of the rather middling bricks we are obliged to use—middling I mean for appearance, because with respect to hardness and soundness they are very good indeed.

E

"I enclose you a sketch of the present state of the viaduct, from which you will be able to judge of our progress.

"We have about 100 bricklayers at work at the viaduct at present, which is as many as can be employed with the present centreing, but I expect to-morrow night the Greatham Beck culvert will be completely finished and the centreing, and thirty men will then be removed to the viaduct.

"We have now only ten piers remaining to be piled, and as there are nine engines to drive them I expect to have them finished next week, and when the foundations are well prepared we shall be able to provide materials for the bricklayers with less difficulty.

"On the whole I consider the viaduct to be now proceeding in a very satisfactory manner, because as many men are employed upon it as can possibly work with advantage.

"The bricks, lime, gravel, sand, timber, and other materials required for the work demand the greatest care and attention, in order that nothing may be wanting to keep every workman constantly supplied, and this I find sometimes so pressing that I have several times been under the necessity of causing Greatham cutting to be stopped a shift, to employ the men and horses in providing what was most urgent.

"I am now able to furnish you with an account of the expense of covering the viaduct with coal-tar.

"To cover 46 square yards:

	s.	d.
Coal-tar . . .	10	0
Coal . . .		10
Time . . .	3	6
	14	4

or 3¾d. per square yard.

Pounded clay, 46 yards, 1 inch thick. 11½ cubic yards at 1s. 6d., 17s. 3d.

or 4½d. per square yard.

But as it would be desirable to cover the tar with 4 or 6 inches of soil, it is probable the expense would be about equal.

"The coal-tar done makes a very complete work by being carried up the spandril walls, and in a heavy shower the water immediately runs to the opening of the pier in the lowest point and flows out scarcely discoloured, and the degree of elasticity obtained will effectually prevent its cracking.

"The covering of the spandril walls makes a much better junction to the arches than could be done if they were covered with clay instead of tar, and it certainly has this advantage over clay, that the brickwork is thereby kept perfectly dry, whereas clay would keep it moist and wet.

"As an instance of the preservative quality of boiled coal-tar, I may mention that a few weeks ago, in repairing the bridge at Sunderland (which I suppose is nearly 100 years old), a piece of Memel plank was taken up which had been covered with it, and was found perfectly sound and hard but quite black, and the tar itself unchanged.

"I have been fortunate enough to meet with a very clever man to manage the boiling and preparing the tar, and I am sure if you saw the five arches now finished with it and the spandril walls covered, you would be pleased.

"Every other part of the line is going on favourably except the laying of the permanent way, for which we have not yet sufficient men; but I have caused advertisements to be issued, which, I have no doubt, will procure them in a few days."

The Stockton and Hartlepool Railway was opened early in 1841. Mr. Fowler, who, as we have already seen, was wont to identify himself with his work in a strongly personal way, seems to have been a little hurt at the absence of his family from the opening ceremony. In a letter, dated March 15th, 1841, and written from Greatham, after dilating on this grievance, he goes on to describe his own connection with the railway.

"The dinner passed over, and the next morning found me early at the Stockton station with a slightly feverish brow and a great many arrangements to make for the traffic and trains, and at eight o'clock I mounted the engine for the first trip to Hartlepool, and kept to my post (on the engine) all day, all the traffic arrangements and conducting of the trains and orders at the stations appearing to devolve on me. I was up a good deal of the night again, and next day was on the engine all day again. This day in performing the last trip we had an accident with the engine by blowing out the lead plug near Hartlepool, which kept us up till between two and three in the morning to repair and get the engine back to Stockton. Since that time I have been nearly as much engaged, but we have had no accident whatever, excepting killing a horse belonging to Hutchinson, the contractor. The accident was to the last train from Hartlepool to Stockton (being dark at the time), and one of Hutchinson's lads was leading bran down the Clarence Railway. On the arrival of the lad at a certain part of the line, he was cautioned not to proceed further until the train had passed, as the Clarence has only a single line at that place in consequence of replacing some old materials; but the fellow said he would go into a siding immediately below and wait till the train had passed. And had he done so all would have been well, but he passed on, and met the train full. The engine driver saw him a short time before they met, and slackened speed as much as possible, but the concussion was sufficient to throw the engine off the rails, kill the horse, scatter the bran, etc.

"I was not on the engine at the time, but attending a parish meeting about a mile distant respecting the railway, and they immediately sent for me. I collected a force of men with screw-jacks and crowbars and winches, and we got her on the line at half-past twelve, and with the other engine dragged her to Stockton."

A large and varied responsibility was thus thrown on the young man of three-and-twenty, and the next letter shows that his capacity and services were obtaining recognition. On August 12th, 1841, he writes to his father from Stockton-on-Tees :—

"I have had Mr. Leather over here to the general meeting, and during his stay several of the directors of the Stockton and Hartlepool Railway took an opportunity of speaking to him about my continuing with them. He replied very handsomely that he had no objection if they wished it, and he advised my continuing to manage the traffic of the railway until Hutchinson was done with his contract, and, of course, when my engagement with Mr. Leather might expire, and then they might make an agreement for my continuing with them. Mr. Leather said he thought I should not be content to remain for less than £300 a year; and they said although well aware they could obtain persons who might perhaps be able to manage for them for less salary, they would have no objection to give that. And there for the present the matter rests. I have fitted up, or am fitting up, an office in the station, and am looking out for lodgings in Stockton."

Towards the close of the letter he adds that he is going to Newcastle about some new engines; apparently he was now entirely responsible for the management, traffic, and locomotive department of the new line, of which he had himself been the constructing engineer. Such a concentration of authority in one person was rare even in those early days, and in after life Sir John Fowler was wont to refer to his connection with the Stockton and Hartlepool Railway as a very valuable experience. Throughout life John Fowler seems to have revelled in hard work, and whether his day's occupation was a difficult problem in engineering or, as here, the routine of a responsible but still commonplace task, his was one of those fortunate natures which require neither recreation nor refreshment apart from work.

Herein no doubt lay one of the secrets of his success and happiness.

There are men who work hard at trying to amuse themselves, and meet with only moderate success. The most enviable life is that which is fully occupied in some congenial and engrossing pursuit. Fortunately many men find their happiness in what is superficially regarded as the drudgery of daily life, and for many years this was Fowler's attitude towards his profession. There are those who complain of the too narrow scope of the industrial life, but the lesson taught by a candid consideration of the facts seems to be, that it is quite possible, by a reasoned submission to the inevitable, to find, both in work successfully accomplished and in the leisure earned thereby, some measure of artistic satisfaction. The career of a successful engineer like Sir John Fowler is to some extent exceptional, but it is still a fair illustration of a wise philosophy of industrial life. A long career of useful and unremitting toil, from which he derived a never-failing enjoyment, brought him ample means and leisure to indulge his love of hospitality and society, sport, travel, and art. The biographer has been struck, in turning over the materials confided to him for the purpose of this work, by noticing how much wider the circle of interests seems to grow in middle and later life, and further, how on the whole, at every period of it, his life seems to have been a happy one; so much so is this the case that, though undoubtedly each addition to wealth and reputation was duly appreciated, the impression is firmly created that even had his ambition been narrower, and his success less marked, Fowler would still have been a happy man. Youth is generally supposed to be the period of wide aspiration. The doors of the prison-house of life tend to close, the poet tells us, on the growing man. This process was, we believe, reversed in the case of John Fowler.

The habit of getting immersed in business was not unnoted by Fowler himself. He does not reprobate it or excuse it, but notes it as inevitable in the following comment on the failure of Mr. J. T. Leather and his brother Henry to visit him according to promise.

"It is," he says, "a most extraordinary circumstance, how people in business acquire the habit of never going to any place unless there is some advantage in it, and I therefore shall never speculate on seeing him, unless his profession calls him here."

Though they were all too busy for mere visits of ceremony or pleasure, the young man was very ready to take much trouble in order to assist his brothers and to advise his father as to their professional prospects.

Early next year, *i.e.* in March, 1843, he was in London on parliamentary business connected with the Hartlepool extensions.

"We are encountering a very fierce opposition on every point, and if we succeed at all it will be by fighting inch by inch."

On May 22nd, 1843, he writes :—

"I am here, as you are aware, much longer than I expected, but the business about which I came is concluded to-day, so far as regards the House of Commons. In due time it will be introduced into the House of Lords, when a similar fight will again have to be encountered.

"In the last week, however, I have been more or less engaged in some other matters, and to-day I have been engaged in the different committee-rooms in three different questions. The Hartlepool Junction Railway, *against* (now finished); the Monkland and Kirkintilloch Railway, *for* (postponed for a week); the Ballochney Railway, *for* (which I have to attend to-morrow)."

This practice is, he remarks, "satisfactory and profitable," and he wishes he had more of it. A cross-examination in a committee-room has now no terrors for him.

In September he is back again at his post on the Stockton and Hartlepool Railway.

"I think it very likely some change will take place here soon, which will affect me, from the traffic falling so very short of their expectations, and it will probably end in its being leased (as in truth it ought to be) to the Clarence Railway, with which it communicates, and of which it ought, when first projected, to have been a branch.

"We are preparing plans for applying to Parliament next year for a dock at Hartlepool, and, from the continued illness of our secretary, and having all his duties and money matters to attend to in addition to everything else, I am at present closely confined, or I would come over to Wadsley some day soon, and ask Henry to meet me.

"I have been very anxious about Henry for some time, from knowing his position, where really he is learning nothing, and from a letter I have lately received from him."

On January 19th, 1844, he writes to his father :—

"We expect very severe opposition to our new docks at Hartlepool, and an immense mass of engineering evidence is expected to be given. I shall at least gain information of great value in the contest, whatever may be the fate of the Bill. Sir Gregory Lewin is retained for us, and he has been with me two days upon the work, getting to understand the engineering and nautical points of the case. I have engineers or contractors with me almost every day, who are expected to give evidence for us, and this takes up much of my time; and every moment I can spare from them I devote to a chart of the coast I have prepared, with soundings, surveys of rocks, eddies, etc., etc., and also the preparing of a model of our present mode of shipping coals at Hartlepool, so that you may conceive I am pretty fully occupied. I expect to be in London in about three weeks. . . .

" I believe this session will be a very busy one in Parliament with railway Bills. I have not got anything except the dock, although I believe I shall be engaged in some other matters. I have been spoken to about a railway in Cornwall by a party who wishes to have my assistance in Parliament."

The Stockton and Hartlepool Railway was a short line of some eight miles. It was subsequently worked in close connection with the Clarence Railway, and the whole combined length amounted to 43 miles. In 1852 the whole was amalgamated with the Hartlepool Docks, then in course of construction, and the title of the undertaking became the West Hartlepool Harbour and Railway Co. Two years later, by the Act of July 31st, 1854, the company was amalgamated in the system of the North Eastern Railway, in the balance-sheet of which there still figure certain West Hartlepool Primary charges representing the un-redeemed portion of the shares of the old Clarence Railway Co.

The Ballochney Railway, to which allusion is made above, was a mineral railway in direct communication with the Slamannan and the Monkland and Kirkin-tilloch lines, and through the latter with the Garnkirk and Glasgow, and also with the Wishaw and Coltness Railways. It was authorised by an Act of Parliament in 1826. Up to 1840 we are told by Whishaw in his *Railways of Great Britain and Ireland*, published in 1840, this line had been worked chiefly by horses, but at that date it was being prepared for locomotive traction. It also contained a self-acting run of 1,200 yards on some modification of the "switchback,"

altogether a very primitive work. The whole of these
lines is now incorporated in the North British system.

Mr. Fowler's inspection of these railways was undertaken through the invitation of Sir John MacNeill, the
well-known Scoto-Hibernian engineer.

The Stockton and Hartlepool Railway was an undertaking of comparatively small importance, but it gave
Mr. Fowler a bit of experience which is probably
unique in the annals of railway engineering. He
seems to have surveyed and then built the line, and
then to have combined in his own person the responsibility for every branch of railway management, from
buying the engines to shutting the carriage doors.
In the course of this employment he picked up a great
deal of miscellaneous information bearing on his profession which is not readily acquired under the minute
subdivision of labour which even then was becoming
usual.

CHAPTER III.

THE RAILWAY ENGINEER

1844-1850

IN the early part of 1844 young Mr. Fowler came to London, and his independent work as an engineer may be dated from that year. In a "List of various works upon which Sir John Fowler has been engaged," kindly furnished from his office, there are three entries only for 1844. Great Grimsby Railway, Sheffield and Lincolnshire Railway, Opposition to Sheffield and Lincolnshire Railway.

In 1845 the list is longer and more imposing: The Great Grimsby Railway, Opposition to Great Grimsby Railway, Sheffield and Lincolnshire Railway, Opposition to Sheffield and Lincolnshire Railway, East Lincolnshire Railway, Great Grimsby and Potteries Railway, Wakefield, Pontefract, and Goole Railway, Sheffield, Barnsley, and Wakefield Railway, Humber Survey, Great Grimsby Works, Wakefield, Brigg, and Goole Railway, Sheffield and Lincolnshire Extension, Great Grimsby Extension, Nos. 1, 2, and 3.

The list for 1846 contains thirty-six entries, and that for 1847 thirty-eight; so that it may be said that at once the young engineer stepped into a large and lucrative practice. Apart from his chief employment in connection with the Lincolnshire railways and docks, there are entries of consultations and work done with regard to railway stations at Sheffield, railway and harbour work at Cockermouth and Workington, waterworks at Sheffield, and a line of direct railway from London to Exeter, and a number of small railways in the neighbourhood of Glasgow; and in the following years it may, without exaggeration, be said that his practice had extended into every part of the kingdom.

In 1851 we find an entry for work done in connection with the Rheims and Douai Railway, and in the same year began his connection with the Oxford, Worcester, and Wolverhampton Railway, now an important section of the Great Western system. This last work had been begun by Brunel, but was at this date transferred to the charge of Mr. Fowler, who carried it to completion.

It would be tedious to narrate in detail the manifold employments in which Mr. Fowler now became engaged. Some of the larger undertakings with which he was associated will be described later on. His task at this date was not only the construction of works, but also the conduct through Parliament of the legislative measures necessary to their inception, amalgamation, and extension. The railway mania, then about to begin, made most urgent demands on the time and services of every capable engineer in the kingdom, and naturally the cool-headed, competent young Yorkshireman was in great request. In after years, discussing his career with a son of his great predecessor and friend, Brunel, Mr. Fowler dwelt on the comparatively prosaic character of much of the engineering work of this period. The era of railway romance came to an end with Stephenson, and Mr. Fowler on the occasion mentioned, with a certain paradoxical exaggeration, described the work of railway construction at this time as pinning down tapes over a large scale map and then cutting them up with a pair of scissors into convenient lengths for construction.

A new departure, in his own career at all events, he always maintained was begun when in 1853 he became associated with the first scheme for tunnelling a railway under the streets of London, an enterprise to which we propose to devote a special chapter. We feel justified, therefore, in passing somewhat lightly over the work of this period and dwelling rather on the general conditions under which the engineer's calling was then conducted.

Much of his work, as already said, was done in the committee-rooms of the Houses of Parliament. Of his views of the duty of an engineer in this respect the best account to be given is that contained in his presidential address to the Institution of Civil Engineers, delivered in 1866.

"All classes of the profession, but especially the railway, the dock and harbour, and the water-works engineer, must possess a knowledge of parliamentary proceeding, so as to be able to avoid all non-compliance with the Standing Orders of Parliament. To do this, it is true, is no easy matter, as the clauses are often drawn up with so little care and practical knowledge that neither engineers nor solicitors, nor the most experienced parliamentary agents can understand what is intended.

" On the subject of parliamentary proceedings generally, it may be taken for granted that all committees desire to do justice to the cases which are brought before them, and that if they sometimes fail in their decisions, either as regards the interests of the public or in arranging a fair settlement between antagonistic interests, it is not unfrequently due to the imperfect and crude manner in which the cases are presented to them. I would therefore impress on all young engineers the importance, both to themselves and to their clients, of laying their cases before committees in the most perfect manner possible, accompanied by full and correct information, carefully prepared and clearly worked out."

The work of an engineer has necessarily to be carried on with the assistance of a public which is in large measure deficient in technical knowledge. Not only parliamentary tribunals, the counsel who practise there, but also directors, shareholders, and the public who may become shareholders, have to be instructed and conciliated. A large measure of Fowler's success was due to his proper appreciation of this fact. He was an admirable witness, and a most persuasive advocate. As we have already seen, he espoused with warmth any cause which he advocated, and naturally identified himself with his clients, and even as a boy he was keenly alive to the mischief done to a cause by the slovenly presentation of it by counsel and witnesses. Many jibes have been made at the expense of the "expert witness," but they are due largely to a misconception of his position. An expert witness, as we all know, is examined on oath, but at the same time his position is very generally recognised as, in some respects and within certain limits, that of an advocate for the side which retains his services. There are limits in advocacy beyond which an honourable man, whether on oath or not, will refuse to go. A railway company, demanding from Parliament compulsory powers, supports its cause by advancing estimates and hypotheses of a highly technical character. It is necessary, owing to the uncertainty which must attend any calculation of this character, that each link in it should be closely scrutinised. Both petitioners and opposers are justified in having the technicalities of their case presented by competent witnesses. There is, naturally enough, a disposition on the part of counsel and litigants to press such witnesses to pass beyond the letter of their brief and to affect a judicial attitude which they ought not to be asked to assume. In discussing with the author the ethical aspects of the question, one of the most distinguished of living engineers admitted the somewhat anomalous conditions, and the embarrassment in which the zeal of counsel on his own side occasionally placed him. As evidence of the versatility of the expert witness, a well-known parliamentary solicitor described to the author his visit to an accomplished engineer for the purpose of securing his assistance for some application to a parliamentary committee. The answer was one of regret that the request had not come earlier, as he had already committed himself to give evidence for the opposition. This readiness to take the responsibility for arranging the technical arguments relevant to the acceptance or rejection of any particular engineering work requiring compulsory powers from Parliament does not necessarily imply any trifling with truth. The legal technicalities of a case are set out by a barrister who is not on oath, while the engineering technicalities, often much more abstruse than the other, are set out by expert witnesses on oath. Among those chiefly concerned the situation is well understood, and, jesting apart, there is not in the mind of reasonable men any suspicion of the integrity of an honourable profession which is obliged to conduct an important section of its business under these peculiar conditions.

The railway mania of 1844–5 found John Fowler fully equipped for the work of railway construction. It carried him forward almost without effort to fortune and reputation; and when, as was inevitable, the reaction came and the high waters of speculative enterprise began to ebb, his abilities retained for him the commanding position which he had already gained in his profession. Some general account of an episode so important to Mr. Fowler's career will not be out of place.

The excitement of the time has been often described. Up to 1844 the railway industry had been pursued with moderation and caution. It provided a great public convenience and a remunerative investment for the savings of all classes. Some of the early companies were declaring good dividends, and there seemed every prospect of continued prosperity. Mr. Francis, in his entertaining *History of the English Railway*, published in 1851, writes as follows :—

"The traffic of the country had trebled within the previous twenty-one years. Three railways, the London and Birmingham, the Grand Junction, and the York and North Midland, paid ten per cent., while a fourth, the Stockton and Darlington, divided fifteen per cent. The safety of the locomotive had also been proved. In 1843, seventy railroads had conveyed 25,000,000 passengers for 330,000,000 miles with only three fatal accidents, and that, too, at an average cost of $1\frac{3}{4}d.$ each person."

The success of the railway as an investment, added to its obvious convenience, and the benefit occasioned by the demand for labour, seemed to justify from every point of view the zeal of the new adventurers.

"It was calculated that were 2,000 miles of the projected roads completed, 500,000 labourers would be employed for four years; that the poor rates must necessarily diminish; that the consumption of excisable liquors would increase, and that the revenue of the country must improve."

The press, the pamphlet, and the pulpit vied in eulogising the new invention. Fine writing, says Mr. Francis, was at a premium, and he proceeds to give a few choice specimens.

"Railways," said one dithyrambic journal, "are the emblems of internal confidence and prosperity. They are the prophetic announcement of an open-eyed people that will not waste their dearest action in the tented field, but exhibit it in the mightier fields of commerce."

The political economist was appealed to by another:—

"Do the people want present employ? Railways give it to hundreds of thousands at this moment. Is it desirable that the artisan or mere labourer should at all times be able to transfer his skill or his strength to the place where he can most profitably employ either? Railways give the power to do so. Is it desirable that prices should be equalised generally through the country? Railways are the great levellers, bringing the producer and consumer into immediate contact. . . . By railways the whole country," piously added the writer, "will, under the blessing of Divine Providence, be cultivated as a garden."

There is more truth perhaps in these sanguine predictions than Mr. Francis, writing in the depressed period of 1851, is prepared to admit; but there is no

F

doubt that, for the moment, expectation as to the immediate commercial success of railways was much exaggerated.

The literature of the prospectus, says Mr. Francis, is worthy of note. One route "would disclose a succession of picturesque scenes," another "traversed a country of unrivalled beauty." The direct London and Exeter, an undertaking in respect of which Mr. Fowler seems to have been consulted, was proposed partly because "it was nearly the road adopted by the Romans."

While the mania lasted, the increase of railway proposals was extraordinary. In a compilation supplied to the *Times* by Mr. Spackman in November, 1845, the national liability in respect of railway extension is set out.* Forty-seven companies had completed their works and had absorbed some $70\frac{1}{2}$ millions of money, of which $22\frac{1}{2}$ millions were raised by loan. There were 118 lines and branches in course of execution, and, of the 67 millions required for their construction, only $6\frac{1}{2}$ millions had been paid up. There were 1,263 railways projected. In respect of these 59 millions had to be deposited as a preliminary to their consideration by Parliament. The total capital paid up, together with the deposits required on new projects, was £113,612,018. The liabilities in the way of loans and expenditure incurred or to be incurred in the construction was £590,447,490. How all this money was to be found no one knew.

There were not wanting warnings that speculation was going too fast. The *Bankers' Magazine* ridiculed the suggestion put forward by some enthusiast that

* See EVANS's *Commercial Crisis of 1847-8*, p. 22.

"we are to have railway streets in London, with carriages overhead, and the foot passengers and shopkeepers underneath; while in the country, railway steam-engines on the atmospheric plan are not only to perform all the work of the lines, but are to employ their surplus power in impregnating the earth with carbonic acid and other gases, so that vegetation may be forced forward despite all the present vicissitudes of the weather, and corn may be made to grow at railway speed."

Other influential persons and journals, notably the *Times*, tried to stem the torrent. Mr. Glyn, the banker, chairman of the London and Birmingham Railway (now the London and North Western), pointed out that railway property by reason of this wild speculation was in great danger. Mr. Hudson, the Railway King, warned the public against an unnecessary and over-hasty multiplication of lines. Mr. Saunders, chairman of the Great Western Railway, spoke in similar terms; but remonstrance was in vain. The " boom," to use a modern term, continued during the spring and summer of 1845, till on October 16th, 1845, as a first sign of the impending reaction, the Bank of England raised the rate of interest.

Before the end of the month a panic, ultimately commensurate with the previous speculation, had begun. Prices reached their lowest point some three years after, namely in October, 1848. Although in many instances not a sod had been cut, and notwithstanding the heavy liability attached to them, the shares of most of these railways were at a premium. When confidence was shaken there ensued a wild anxiety to " get out." Many subscribers had applied for more stock than they could pay for, partly because they saw that allotments in full were not made in the early stages of the mania,

and partly because premium-hunting had been hitherto
a remunerative pastime. Now, the difficulty was to
find the allottees of the shares and to extract from
them the balance of liability on their allotments. "On
one projected line," says Mr. Francis, "only £60 out
£700,000 was realised by a call. No other panic,"
says this well-informed authority, "was ever so fatal
to the middle class." The following statement of
prices will show as clearly as anything else the destructive effect of the panic.

	AUGUST, 1845.			OCTOBER, 1848.	
	Amount of share.	Amount paid up.	Highest price.	Amount paid up.	Highest price.
	£	£	£	£	£
Brighton . . .	50	... 50	... 80½	50	... 29
Caledonian . .	50	... 5	... 12⅝	50	... 20¾
Eastern Counties .	25	... 14/16/0	... 21¼	20	... 13⅝
Great Western . .	100	... 80	... 236	90	... 80
L. and N. Western .	stock	... —	... 254	stock	... 121
Midland . . .	stock	... —	... 183	stock	... 86
South Eastern . .	av. 33/2/4 ...	—	... 48½	av. 33/2/4	... 24¼
South Western . .	av. 41/6/10 ...	—	... 84	stock	... 42
York and N. Midland	50	... 50	... 112	stock	... 54

The Board of Trade on November 28th, 1844, had
announced that, in virtue of the authority conferred
on it by the legislature, it would require evidence on
certain points from promoters before reporting to Parliament in favour of their proposals. On December
31st a report was issued by the Board which, if its
policy had been adopted, would have restricted competition within what the Board conceived to be legitimate
limits. Parliament however set this advice on one side,
and adopted the principle that a railway had no vested
interest, and was not entitled to be protected from
competition.

Mr. Morrison urged in Parliament the necessity of

regulating the new industry, but the result is thus summed up in a passage quoted by Mr. Francis.

"Swayed by motives which it is difficult to fathom, the two Houses, with singular unanimity, agreed to reverse their wise decisions (*i.e.* the recommendations of the Board of Trade), and to give unrestricted scope to competition. Little regard was paid to the claims and interests of existing railway companies, still less to the interests of the unfortunate persons who were induced to embark in the new projects for no better reason than that they had been sanctioned by Parliament. The opportunity of confining the exceptional gauge within its original territory was also for ever thrown away. By an inconceivable want of statesmanlike views and foresight, no effort was made to connect the isolated railways which then existed into one great and combined system, in the form in which they would be most subservient to the wants of the community, and to the great ends of domestic government and national defence. Further, the sudden change from one extreme of determined rejection or dilatory acquiescence to the opposite extreme of unlimited concession gave a powerful stimulus to the spirit of speculation, and turned nearly the whole nation into gamblers."

Such was one line of argument, and from it we may formulate the following statement of grievances :—

1. That Parliament had not protected railway shareholders or endeavoured to secure existing railway companies from competition.

2. That it had not enforced one gauge for the whole railway system.

3. That it had allowed railways to be isolated, whereby the interests of "domestic government and national defence" had been neglected.

4. That by giving unrestricted scope to competition it had stimulated speculation.

The indictment is an interesting one, if only because it shows how difficulties of this character often succeed in solving themselves. No body of railway controversialists now ask to have special legislation in favour of shareholders. The gauge question, after experiment, has solved itself without legislative interference. There is, as far as we are aware, no demand for military railway lines in this country. A large proportion of railway property is held by trustees and persons who select it because it is not a speculative investment. Even the margin of ordinary stock is not speculative in the popular sense of the term. The neglect of Parliament, as conceived by the above-quoted complainants, does not appear, therefore, to have had very serious consequences.

There was of course another side to the shield. If the party which desired a stricter regulation of railway enterprise was disappointed, the railway interest itself had grounds of complaint. Mr. Robert Stephenson, in his presidential address to the Institution of Civil Engineers in 1856, gave expression to this feeling. He pointed out the illogical position of a legislature which assumed that, owing to the facilities for combination, competition was impossible between railways, and yet went on authorising duplication of lines to the ruin of shareholders, and, insomuch as the new competition was invariably terminated after a period by fresh combination, without any advantage to the public. By this policy, as he puts it, the capital of railway enterprise is constantly being increased, while the traffic is being divided.

Robert Stephenson's often quoted aphorism, which he on this occasion imputes to the legislature, that " where combination is possible competition is impossible," may suggest to some minds the corollary that a railway company can and does charge exorbitant rates. This was not Stephenson's own view, as the following, which he himself with great deliberation has inserted in this same presidential address, will show.

"It may be thought," he says, "that with respect to fares, the interests of railway companies and of the public are antagonistic. Regarding the question, however, with a more enlarged view, it will be readily seen that so far from those interests being opposed, they are, in all respects, identical. Fares should be regulated by directorates, exclusively by a consideration of the circumstances which produce the largest revenue to the companies, and the circumstances which produce the largest revenue are those which induce most travellers to avail themselves of railway facilities. As regards the public, it may easily be shown that nothing is so desirable for their interests as to take advantage of all the opportunities afforded by railways. As regards railways, it is certain that nothing is so profitable, because nothing is so cheaply transported, as passenger traffic.* Goods traffic of whatsoever description

---

* Adam Smith (*Wealth of Nations*, book i. chap. viii.), living before the age of railways, said that "a man is, of all sorts of luggage, the most difficult to be transported." Nothing marks the extent of the economic revolution produced by railways more clearly than the complete contradiction that may now be given to Adam Smith's proposition. Railway communication permits men (and the permission is now being brought to the reach of the manual labourer) to live in one place and do their daily work in another, and to move from place to place where the "best stock or employment is to be found." It has thus gone far to release the labouring class from the immobility which at one time tied them to unremunerative toil upon the land of the parish where they were settled, thereby adding enormously to the efficiency of labour. It is hardly too much to say that, under this newly acquired mobility, the units of labour are acquiring something of the character of interchangeable parts of machinery, a principle of the most far-reaching importance. When it is added that the motive power, in the circulation of labour, is the attraction of better remuneration or improved conditions of toil, and the

must be more or less costly. Every article conveyed by railways requires handling and conveyance beyond the limit of the railway-station; but passengers take care of themselves, and find their own way at their own cost from the terminus at which they are set down."

This passage sets out very correctly the grounds on which confidence is claimed for the principle of freedom of enterprise. The term competition as ordinarily understood is a very inadequate description of the guarantee for public convenience which we believe is contained in that policy. The influence which has prevented railways from being an oppressive monopoly has not been merely the competition of other modes of transport, (though this competition has not been wanting), nor the regulation of authority, but the much more effective competition of the natural tendency of men and things to remain where they are. Immobility is cheap and simple, and is far the most formidable competitor with which railway companies have to contend, and it can only be overcome by increasing the cheapness, facility, and comfort of transport.

To return, however, to the mania of 1845.

The incipient signs of panic in the early autumn of 1845 did not interfere with the deposit of plans which, according to the regulations of the Board of Trade, had to be made on or before the 30th November of that year.

result, as regards commodities, greater efficiency and cheapness of production, we have gone near to formulating the very essential principle of modern civilisation—high remuneration for labour and the rapid multiplication and increased cheapness of commodities. We are able herein to see how important a part has been played by that most valuable implement of commerce, our modern railway system.

"The 30th November, 1845, the day by which the documents were to be lodged, fell on a Sunday, but there was no Sabbath for the restless railway promoter. 'The stir of agents,' says the *Railway Chronicle*, 'made Sunday anything but a day of rest or devout observance throughout the country. The offices of clerks of the peace and the doors of the Board of Trade were stormed by breathless depositors till the stroke of midnight. Frantic "standing-order missionaries" from Harwich, arriving a few minutes afterwards — miscarried, alas! by blundering postboys, who drive for an hour and a half about Pimlico seeking the office in vain—have to besiege its inexorable doors and fling their plans into the lobby, breaking the passage lamp, with no effect but that of having them flung back again in their doleful faces. . . . On the Great Western Railway the haste to overtake spare minutes had nearly led to a tragedy dark enough to fill the courts of Gray's Inn and the purlieus of Chancery Lane with inconsolable mourning. A squadron of solicitors to some of the projected lines had borrowed the wings of an express, which unhappily broke down at Maidenhead. In this disabled condition the engine was charged by another, which had started with several legal gentlemen connected with the Great Western and Exeter Companies, and the carriage with the learned freight was dashed to pieces. . . . The scared pursuivants shook themselves, packed up their ruffled plans, charitably picked up the stranded attorneys, whose wreck had nearly caused a dismal hiatus in the profession, and heroically steamed onwards, arriving, we are glad to hear, in good time. . . . A collision between engines on the broad gauge we take to be as smart an encounter as any tilting encounter. . . . On the Great Western Railway on Sunday there were ten express trains similarly employed'; and reading this, we deem it a great mercy that we have no worse casualty than the above record."*

The plans had to be deposited not only at the Board of Trade, but at the offices of the Clerk of the

* FRANCIS, *History of English Railways*, vol. ii. p. 248.

Peace in the several counties. The Clerk of the Peace at Preston kept his office shut on the Sunday, considering that the order as to Sunday opening applied to the Board of Trade only. The railway promoters took a different view, and stormed the office, and flung their plans in through the broken windows.

The Eastern Counties Company ran eighteen or twenty special trains for their various projected lines.

"The majority of the plans from the provinces," says the *Morning Chronicle*, "have been sent up by express trains, and it is whispered that those companies with the locomotives at their command, and to whom the lines belonged, availed themselves of this advantage to such an extent for the exclusive transmission of their own plans and sections, as actually to refuse special trains to their competitors."

One competing company was driven to the stratagem of enclosing their plans and the clerk in charge in a hearse in order to obtain a special train from a rival company.

Curious illustration of the degree to which railway enterprise monopolised the services of the engineering profession may be furnished from both ends of the scale. That competent engineers were fully employed is natural enough, but so deep were their engagements, and so satisfactory to them their terms of remuneration, that when in 1845 Mr. James Morrison, M.P., urged on the Government the propriety of holding an inquiry conducted by some competent and expert authority, Lord Dalhousie, who at this time was at the head of the Railway Department of the Board of Trade, took refuge in the excuse "that he really could not find a person competent from his qualifications for such a commission, as the railway companies had engrossed all the talent available for the purpose."* Mr. Brunel was said to be connected with fourteen lines, Mr. Robert Stephenson with thirty-four, Mr. Locke with thirty-one, Mr. Vignoles with twenty-two, Sir John Rennie with twenty, and Mr. Rastrick with seventeen.

At the other end of the scale the demand was equally fierce. Mr. Williams, in his excellent volume, *Our Iron Roads*, describes the demand which arose for even the humblest service in the following terms :—

"Innumerable surveyors and levellers were required, and in many instances they made from six to fifteen guineas a day ; while numbers of persons were employed who were acquainted with only the rudiments of the art, and who, by their blunders, subsequently occasioned even fatal inconvenience to the enterprises in which they were concerned. The extravagant payment that was offered also induced great numbers to leave situations they occupied in order to learn the new business ; while professors, lecturers, and teachers announced classes, lectures, and private instruction which, with almost magical celerity, would convert all persons of ordinary powers into practical men, earning enormous payments. Still the demand was not equal to the wants of the case. Surveyors and levellers became worth their weight in gold, and countless amateurs presented themselves. A peddling stationer, who long itinerated in Northumberland and Durham, earned 'five guineas a day and his expenses' on a southern railroad ; and the *Lancaster Guardian* stated that a fat neighbour, long unemployed, obtained an engagement of three guineas. 'I could have had five,' said he, 'but it would have been in a county where the gradients were severe and too trying for my wind'; and he preferred three guineas

* *The Influence of English Railway Legislation on Trade and Industry*, by JAMES MORRISON (1848), p. 25.

and a level line. No fewer than eighty surveyors arrived in Lancaster in one day for the York and Lancaster line only, and they were followed by another batch a few days afterwards."

No record of strange adventures has been preserved as to the persons and plans that raced to the Board of Trade on that eventful Sunday on behalf of the lines with which Mr. Fowler was concerned, but without doubt they were all involved in the general turmoil.

The following passage of a letter, dated 5th March, 1890, from Mr. Bernard Wake, recalled to Sir John Fowler some of the incidents of that time.

"What strides the world has made since the 1844 days, when the Sheffield and Lincolnshire Railway was launched. You and I, with divers others, were busy then at Brigg for days together, without going to bed; and eyes got red and hearts got heavy when plans did not arrive from London. Carriages were at the door to deposit them, but the clock stole on, time was up, and plans were lacking, and the Sheffield and Lincolnshire scheme appeared to be 'as dead as a door nail.' You will recall those early days of railways. I well remember travelling from Brigg fast asleep the whole way to Sheffield with R. I. Gainsford. He, when we got to Sheffield, thought we were at the first toll-bar out of Brigg. . . . We had been sleepless for days, only catching bits of sleep whilst examining books of reference, especially R. I. Gainsford, who in reading, half asleep, mistook his own finger-end for a spider running over the paper!

"Then one recalls, in 1845, 'The Great Grimsby, Sheffield, the Potteries, and Grand Junction Railway,' otherwise 'The Great Grouse, Trout, Ling-besom, and Billberry Junction Railway,' otherwise 'The Flute Railway'; and how the Derbyshire hills were to be pierced years ago, if that scheme had been acceptable to Parliament."

In after life Fowler was fond of telling a story which relates to this period; how an excited railway promoter arrived at dead of night in a coach-and-four at his father's house of Wadsley Hall, where the young engineer happened to be. Mr. Fowler was roused from sleep, and found that his visitor wished him to undertake the engineering of a line from Leeds to Glasgow, and had brought an order for £20,000 as a payment on account of survey expenses. It wanted only a few weeks of the time before the day of depositing the plan. Mr. Fowler prudently declined what, to a young man, must have been a tempting offer, and the coach-and-four drove off into the night.

His main work for the next few years was the laying out of the group of railways then being promoted from Sheffield—the Sheffield and Lincolnshire, the Great Grimsby, the New Holland, the East Lincolnshire, and others, which finally became amalgamated under the title of the Manchester, Sheffield, and Lincolnshire. For the construction of these Mr. Fowler was the chief engineer, and on him fell the responsibility of carrying the necessary legislation through Parliament.

Among the papers relating to this period are the memoranda of "Information prepared for the Railway Department of the Board of Trade, in conformity with their circular of November 28th, 1844," with respect to the Sheffield and Lincolnshire Junction Railway, and also with respect to the Great Grimsby and Sheffield Junction Railways. The memorandum with regard to the last line records *inter alia* that at a meeting at Great Grimsby on November 1st, 1844 (the Right Hon. the Earl of Yarborough in the chair), the following resolution, proposed by Edward Heneage, M.P. for Grimsby, seconded by T. G. Cobbett, Esq., of Helsham, was unanimously carried :—

"That this meeting, having considered the report of John Fowler, Esq., the engineer of the railway company, is of opinion that the route recommended by him *viâ* Glandford Brigg is the best that can be adopted for accomplishing the objects in view, and is entitled to the support of the landed proprietors of North Lincolnshire."

Both documents attest the comparative simplicity and ease of the engineering of the lines, and the great benefit which they would give as part of a trunk line between Manchester, Sheffield, and the eastern ports.

The Great Grimsby line was a favourite at one time with speculators. Everyone gambled in shares, and, as in all gambling, choice was determined by the merest trifles. If a line were fortunate, promoters would endeavour to appropriate as much of its name as they could for other lines, in the hope that their particular venture would gain by the association. As an instance, Mr. Fowler's Great Grimsby Railway was at a premium, and consequently we are told the name of Great Grimsby was frequently brought in quite irrespective of geographical facts. This was done to such an extent that the Earl of Devon, at that time chairman of the railway committee, on one occasion was heard to exclaim, "What! Great Grimsby again! Go it, Great Grimsby!"

The Manchester, Sheffield, and Lincolnshire (now the Great Central Railway), in the construction of which Mr. Fowler was largely employed, consisted, in 1846, of an amalgamation of the Sheffield, Ashton-under-Lyne and Manchester, the Great Grimsby and Sheffield Junction, the Sheffield and Lincolnshire, the Sheffield and Lincolnshire Extension, and the Great Grimsby Dock Companies. In 1847 a further amalgamation was made with the Manchester and Lincoln Union

Company. The whole was then dissolved, and incorporated as the Manchester, Sheffield, and Lincolnshire, by the Consolidation Act of 1849. The subsequent extension of the line, and its recent successful invasion of London as the Great Central, are matters of history. During the long period when it was content to be a junior partner to one or other of the great trunk lines, its ambition was more restricted. The original conception seems to have been the formation of a trunk line running from east to west. The traffic to London, however, has a tendency to deflect all "trunk" operations to the south, and when the Manchester, Sheffield, and Lincolnshire Railway met in its course the Great Northern Railway, there naturally came to be some exchange of running powers. The tasting of the traffic to the "Great Wen," as Cobbett used to call our ever-expanding Metropolitan area, whetted the appetite of the Sheffield managers, who, after nearly half a century of gradual advance, have at length pushed their way into the position of a first-class but somewhat impoverished railway, with a terminus in London.

The facts, though they are only pertinent here as the later history of a line with the early portions of which Mr. Fowler was concerned, are thus given in Mr. Grinling's *History of the Great Northern Railway* :—

"On March 1st, 1848, the Great Northern Railway Company began its first work as public carriers on a length of about 30 miles of railway from Louth to New Holland-on-Humber (opposite Hull). Nearly half this route, namely the 14 miles from Louth to Great Grimsby, was completely under Great Northern control by lease from the East Lincolnshire Company, and the remainder was worked over by arrangement with its owners, the Manchester, Sheffield and Lincolnshire, who in return worked over the East Lincolnshire to Louth."

Mr. Cubitt, the superintending engineer of the Great
Northern Railway, unfortunately died at this juncture,
and his place was filled by Mr. Edward Bury.

"Under his superintendence and that of Mr. (now Sir)
John Fowler, who was engineer both to the East Lincolnshire
and to the Manchester, Sheffield, and Lincolnshire (eastern
section), the traffic from Louth to New Holland was efficiently
conducted."

The Great Northern Railway was at this time, 1848,
a keen competitor in the race to the North. "King"
Hudson was still managing the interests of the Midland
Railway, which as yet had no access to the Metropolis.
A proposal made by Hudson to extend the Midland
from Leicester to Hitchin and there to make a junction
with the Great Eastern Railway, had been successfully
resisted by the Great Northern. The position of the
Great Northern, as its chairman, Mr. Denison, pointed
out in February, 1848, was practically secure. It must
form the southern portion of the East Coast line between
London and the North. Its main line, between York
and Peterborough, had, however, still to be made, and
the Company, owing to the stringency of the money
market, was confining itself to pushing on a circuitous
bit of loop line round some of the Lincolnshire towns.
A Deviation Bill, which the company was promoting
in regard to this loop connection with the North, was
unexpectedly thrown out, and their strong position
was, for the moment, jeopardised. The situation is
thus explained by the historian of the Great Northern
Railway :—

"This (the rejection of the Deviation Bill) was an alarming
event for the Great Northern Board, for it threatened completely to upset the plans which had led them to make the

loop before the towns line; but fortunately they found a fairly satisfactory makeshift in an alliance with the Manchester, Sheffield, and Lincolnshire. This company, as we know, had powers for a branch from its main line at Clarborough, about five miles east of Retford, to join the Great Northern loop at Saxelby, from which point the Great Northern had granted it running rights into Lincoln in return for similar rights over the Manchester, Sheffield, and Lincolnshire from Retford to Sheffield; and although the Sheffield directors (whose chairman, Lord Yarborough, was an old London and York committeeman) had intended postponing the construction of this branch till better times, they now acceded to the Great Northern proposal that the line should be made at once, so that the Great Northern might use it for through traffic in place of the just rejected Gainsborough to Rossington line. Accordingly Mr. John Fowler was instructed by the Sheffield Board to put the work immediately in hand, while at the same time, and with the same object, the Great Northern directors let a further contract to Messrs. Peto and Betts for the section of their main line from Retford to Doncaster."

This alliance with the Great Northern Railway put the Lincolnshire lines and the ports of Boston, Grimsby, and Hull in direct communication with London as well as with Lancashire.

In April, 1849, came the collapse of the Railway King. Hudson resigned the chairmanship of the Midland, which for the moment subsided into the position of a second-class line, and abandoned the competition with the Great Northern Railway for the access to London. The fact largely increased the importance of the Manchester, Sheffield, and Lincolnshire. An alliance at Retford between the Great Northern Railway and the Manchester, Sheffield, and Lincolnshire lines provided an alternative route to Manchester,

G

and threatened a formidable competition to the London and North Western Railway. The London and North Western Railway therefore became anxious to conciliate the smaller company, and, for a period, the Manchester, Sheffield, and Lincolnshire line was detached from its alliance with the Great Northern Railway. Under the guidance of Mr. Allport, at this time general manager of the Manchester, Sheffield, and Lincolnshire, the facilities given to the Great Northern traffic were withdrawn, and the running powers made as inconvenient as possible.

These alliances, ruptures, and reprisals did not bring much prosperity to the Manchester, Sheffield, and Lincolnshire rail, a line whose ambitions have always seemed to be in excess of its resources. In 1853 a committee was appointed with a view of inaugurating a policy of retrenchment, and Mr. (afterwards Sir) Edward Watkin was appointed manager. Sir Edward's subsequent conduct of the railway into many ambitious, and ultimately successful enterprises, does not quite fit in with that limitation of expense which was one, at least, of the original objects of his appointment.

The relations between Mr. Watkin and Mr. Fowler were at first cordial, but the affairs of the line hardly gave scope enough for two men of very masterful character, and before long considerable friction was set up between them, and Mr. Fowler's connection with the railway gradually ceased.

The Manchester, Sheffield, and Lincolnshire continued its alliance with the West Coast confederacy till 1857, when another reconstruction of interests was made—the beginning of the so-called fifty years' agreement between the Great Northern and the Sheffield Companies. In 1861 a breach of this agreement, as it was regarded by the Great Northern Railway, was made by the Sheffield directors, who entered into an alliance with the Midland, then recovering itself from its earlier misfortunes.

It seems, indeed, to have been the policy of this line to play off the different trunk lines one against the other. This alliance with the Midland accentuated the rivalry between the Midland and the Great Northern Railways, and, when the Great Northern withdrew certain facilities granted to it at King's Cross for coal traffic, the Midland thought itself forced to push on and obtain independent access to London— action which resulted in the Midland's successful extension to St. Pancras in 1868.

Many years later, in 1880, Mr. Fowler was concerned in an incident which proved of considerable importance in the tangled history of railway politics in connection with the Manchester, Sheffield, and Lincolnshire Railway. The inhabitants of Huddersfield were anxious to obtain a connection with the Great Northern Railway more direct than the existing one which ran over a portion of the Lancashire, Yorkshire, and the Manchester Sheffield route, and, as a necessary part of this Bill, applied for compulsory powers over a portion of the Great Northern Railway. This railway, as joint owner of the West Riding and Grimsby line, felt bound to oppose. The Manchester, Sheffield, and Lincolnshire did not appear until Mr. Fowler, who with Mr. Fraser was acting for the promoters, and was also consulting engineer to the Great Northern Railway, was put into the witness-box to announce that a compromise had been arrived at between the directors of the Great Northern Railway and the promoters of the Huddersfield connection. The Manchester, Sheffield, and Lincolnshire Railway thereupon instructed counsel to appear and protest against the agreement which had been accepted by its partner. As a result of this, and other opposition, the Bill, or rather the compromise founded thereon, was lost.

This episode led to an attack by the Huddersfield interest on the agreement then subsisting between the Great Northern Railway and the Manchester, Sheffield, and Lincolnshire.* Notwithstanding their opposition, however, the so-called fifty years' agreement was sanctioned by the Railway Commission for another period of ten years. The alliance continued in force with intervals of more or less animated misunderstanding till, in 1892, the Manchester, Sheffield, and Lincolnshire obtained independent access to London under the title of the Great Central. All this is an anticipation of events, introduced here to show how those early labours of Mr. Fowler led to great and important extensions of enterprise.

We shall have abundant opportunity later to describe some of Fowler's engineering triumphs. We dwell here rather on some of the social and economic results which followed, or seemed likely to follow, on the opening of this new country to railway communications. The business of railway construction had now entered on its prosaic period, and the work of John Fowler did not differ from that of many of his compeers. It was characterised, we can imagine, by its thoroughness and common sense. He was, it has always been said

* GRINLING, p. 339.

of him, a good shareholder's engineer. He himself has described the policy which he pursued.

"It is not the business of an engineer to build a fine bridge or to construct a magnificent engineering work for the purpose of displaying his professional attainments, but, whatever the temptation may be, his duty is to accomplish the end and aim of his employers by such works and such means as are, on the whole, the best and most economically adapted for the purpose."

The self-effacement of the engineer who sees that his business is to make a railway, and not to raise a monument to his own ingenuity, deserves recognition. The railway system of this country is a monument not to any single reputation, but to the common sense and business capacity of a great army of railway constructors. The success of a work at this period is to be measured rather by the comparatively commonplace and simple character of its construction than by its boldness and ingenuity. The demand, in railway work, for those gigantic and difficult undertakings which strike the imagination had not as yet arisen, and was not yet justified. Naturally, the easier gradients and the less venturesome experiments were attempted first. It was the engineer's business to avoid rather than to seek difficult and adventurous expedients.

The interest of the work of this period is therefore rather for the economist than for the engineer. It had hitherto been a complaint that the agricultural interest had been neglected by the railway, but this group of railways made a new departure, and became instrumental in connecting the manufacturing Midland towns with the rural districts of Lincolnshire, and with important harbours on the eastern coast.

Mr. Samuel Sidney, the ingenious author of the *Railway System*, a work published in 1846, states the position in the following terms. He observed

"that, while in purely agricultural districts produce sells at barely remunerative prices, while timber rots in the woods for want of a market, and the peasantry shiver through the winter (for want of better fire than faggots gathered by their landlord's sufferance afford), from one and the same cause—the absence of means of conveyance to a market—rich coal mines remain unopened, and there is in the manufacturing districts a much greater demand for agricultural produce than can be easily satisfied. The demand for a luxury affords some idea of the demand for absolute necessaries. When we learn from Captain Lawes' evidence that Manchester consumes in the season from ten to twelve tons of cucumbers a week, we can form some idea of the enormous demand for bread, meat, butter, cheese, and poultry. . . . Our agriculturalists have just received, or imagine they have received, a heavy blow and deep discouragement. They fear low prices, and the cries of alarm are loudest and deepest from those most distant from the vast increasing and unceasing demand for all that farms produce, made by the great Metropolis and the manufacturing counties. There has been a talk of compensation for the agricultural interest. The best, the only real compensation, will be found in extending the farmers' markets and giving him thousands for customers where he had only hundreds before for his purchasers, the range of cities instead of the neighbouring village. This can only be done by an extension of the railway system, and that on such a scale that it shall be worth the while of the company to attend to agricultural traffic. At present the parish roads all converge towards the highways and the nearest town or village; in future these roads must be turned toward the railway station. We shall then see, on market days, farmers proceeding with their stock of produce to a roadside railway station, provided as the focus of a circle of farms. To the farmer to whom the choice will be open of sending his bullocks or butter, cheese or corn, to London or to Lancashire, as he pleases—who can join with his neighbours in having a cargo of bones, oilcake, or guano, fresh from the ship side, who can buy the draining tiles, timber, iron, limestone, and coal wholesale, where formerly he bought it in retail—a new existence, a new course of profit will be opened. It is to such compensation as this that he must look to replace the protection that he lost the other day."

Two years later, Mr. Sidney rode through North Lincolnshire, and published, somewhat after the manner of Arthur Young, a small volume, entitled, *Rough Notes of a Ride over the Track of the Manchester, Sheffield, Lincolnshire, and other Railways,* 1848. His object, he says, was

" to notify to the agricultural and railway world what railways could do for Lincolnshire and what Lincolnshire could do for railways cheaply constructed and liberally worked. I took Lincolnshire for an example of what the like connection might effect in every other agricultural county, where, as in the case of the East Lincolnshire and the Manchester, Sheffield, and Lincolnshire, the companies and the engineer had determined to make their lines a useful 'implement of commerce' for the accommodation of all classes."

Events have moved more slowly, and perhaps not exactly on the lines predicted by this acute observer, but the description which he gives of the Lincolnshire of that day, and of the expectations of an enlightened advocate of railways with regard to this corner of England where it was Mr. Fowler's fortune to be largely employed, warrant us in reproducing some of his remarks.

The phrase of the engineer (who is not named by Mr. Sidney) that the railway is to be " a useful

implement of commerce" is at once happy and accurate.
Under the new era of free trade, the future development of a country or district was to depend on the
ability of its inhabitants to organise an industry
such as would enable them to sell their products in,
and to draw their supplies from, the open markets of
the commercial world. This, it was fully expected,
might bring about a revolution in the industrial
employments of the people. As regards English
agriculturalists, who were not likely to become ex-
porters, the advent of foreign supplies in the home
market seemed a dangerous source of competition.
English land and climate, however, is not so poor, and
the English farmer is not so devoid of ingenuity that
English agriculture need be driven from the markets
which lie at his gate.

In the passage already quoted, Mr. Sidney indicates
the fallacy of supposing that before the era of the
newly-increased facilities for exchange the labourer
enjoyed a golden age of comfort and prosperity. A
few years earlier the whole labouring population of the
land had been partially supported by the poor rate.
The parochial system, a survival in pseudo-philan-
thropic guise of a repressive feudalism, confined the
labourer by a delusive guarantee of poor-law maintenance to the place of his settlement. The railway
system, first by creating an entirely new industry, and
so giving employment to many, and then by promoting
a migration from districts where labour was congested
to industrial centres, where remunerative wages were
to be earned, was in itself a useful implement of
commerce, which, even if, or may we not say because,
it allowed certain old and badly remunerative indus-

tries to decay, was on the whole to the advantage of mankind.   As Mr. Sidney goes on to point out, the railway opened to the agriculturalist new markets for his old products, and suggested to him new forms of enterprise.   In return, he became a purchaser of fuel and other products from the manufacturing districts, which added to his comfort, and also of manures and agricultural machinery, which added to the efficiency of his labour, and to the capital value of the land.

In early railway expansion the growth of goods traffic seems to have been generally much below expectation, and the success of the new movement was assured rather by the volume of passenger traffic, which proved much in excess of calculation.   The arrangements even at the terminal stations were not sufficiently organised to allow of the successful distribution of large quantities of agricultural produce. This, it may be said, is a distinct branch of the great industry of distribution, and required, and indeed still requires, much time and ingenuity before it is put on a satisfactory footing.

The group of railways now under consideration seems to have given, from the first, much attention to this important extension of their business.   Mr. Hudson, the Railway King, has been credited with the merit of inducing the Eastern Counties Railways to make liberal additions to their stations in the way of pens and cattle sheds and warehouses for grain.

"Already," says Mr. Sidney, "much has been done.   For instance, we know that on the eastern coast the straw yards (where that manure is made which is one of the main principles of successful cultivation) are filled by lean cattle imported from Devonshire, Herefordshire, Yorkshire, and Scotland, and even Ireland. These beasts formerly travelled at the rate of 10 miles a day along the roads under the care of drivers. They are now despatched by rail in as many hours as days by the road system, one man doing the duty of five, without dogs."

He goes on to tell of 400 beasts conveyed within twenty-four hours from the moors of Cumberland to the turnip farms of Norfolk without the loss of a beast or a pound of flesh, and remarks that the same beasts, fattened, would probably find their way to Smithfield. "This," he adds, "may be taken as a sample of many hundred unchronicled transactions of the same kind every year." On the shifting of the course of commerce occasioned by railways in this neighbourhood, he makes some interesting comment. The forest district of Nottingham and North Lincolnshire formerly had a profitable trade, conducted without the aid of a railway, with the clothing population of Yorkshire and the town of Manchester, until the easy conveyance between Manchester, Sheffield, York, and Hull, and other towns diverted their best customers to markets within railroad reach. In the evidence given before the committee on the Manchester, Sheffield, and Lincolnshire Railways it was stated that Worksop and East Retford had both lost trade owing to the competitors which the new railway system had brought into markets of which they formerly had a monopoly. These small and temporary inconveniences, he justly remarks, are definite and easily formulated, while it is difficult to give any statistical account of the innumerable advantages which railway communications have conferred on all classes. Mr. Smith of Deanston, a celebrated agricultural authority of the day, giving evidence before a committee of the House of Commons in 1846, calculated that on a typical farm the result of the cheapness and facility of transport produced by railway communications would amount to an advantage to the farmer of about 10s. an acre. He further entertained the opinion that railways would be found useful "for the purpose of mixing soils," and he seemed to contemplate even a larger use of the railway in this respect than the transport of manure and lime and sand, which is an everyday occurrence at the present time.

With regard to the ferries and docks which were a special feature in this particular group of railways, Mr. Sidney takes a liberal and sensible view.

"Parliament," he says, "has lately displayed great jealousy with regard to granting any powers beyond carrying powers to railway companies. Objections have successfully been urged to their becoming owners of docks, ferries, steamboats, and other appliances for assisting or increasing traffic beyond the limits of railway termini. In the only instances in which railway companies have obtained such powers, the public have obtained better docks and steamboats than they would otherwise have enjoyed. The result of what certain ignorant M.P.'s., misled by sounds, have called a monopoly, has been superior accommodation for the public at a cheaper rate. I do not think there would be any difficulty in proving that the public would gain practically, especially the agricultural public, by permitting every railway company terminating in a seaport town to have, not only access to, but the whole control and possession of a dock and warehouses. A railway company of large capital would be able to disregard any profits on, say, half a million, sunk in docks and warehousing in consideration of the large additional traffic which would be attracted by a liberal style of conducting business.

"In bones and other foreign tillages, with the exception of guano, a considerable proportion of the price is composed of the cost of transfer and retransfer. For instance, bones from North Germany are carted from the shipside to a warehouse, and from thence to a railway station. Every day our farmers, as they progress in scientific farming, are more dependent on these importations of foreign tillages. In the hands of a railway company, bones and many other matters would be moved from the ship to the warehouse direct, and, frequently, under an arrangement with the importer, from the ship to trucks which, without halt, would convey the whole cargo direct to some inland depôt at a saving which may be calculated from the fact that the cost of carting a puncheon of rum from St. Katherine's to a railway station exceeds the whole freight from London to Rotterdam. A grocer, before the committee on the Lincolnshire Railways, proved that it was cheaper to send a hogshead of sugar from the London Docks by sea, round to Gainsborough and thence to Sheffield, than to cart it to the Camden Town station of the London and North Western Railway. It is, of course, on articles of low value that small savings in the cost of conveyance tell.

"In certain situations, steam mills and abattoirs might advantageously form part of the scheme of railway stations, even if let off to tenants. Killing beasts, like burying human beings within the limits of towns, will soon, it is to be hoped, be rendered penal. The next step will be to make in each important town a railway beast market and slaughter-house. These companies are corporations formed to our hands to secure orderly responsible management. . . . Certain it is that our railway powers will never afford full advantage to the public until our legislators look to practical results—not to the question of whether powers of building, boating, damming, and docking are contrary to precedent, but whether they will produce good accommodation at a lower rate than previous arrangements."

After a fierce struggle the necessary powers for constructing docks and other adjuncts to the railway system were obtained. Mr. Sidney thus describes the works which in 1848 were proceeding at New Holland :—

"This New Holland, two years ago almost a solitude, was formerly a famous resort for smugglers of Hollands and prime tobacco, of late renown, by a ferry to Hull, which with difficulty supported one coach and one poor alehouse. But Steam, the great magician of the nineteenth century, had been at work, and raised monuments of his deeds on all sides. Hundreds of workmen were engaged in putting the finishing stroke to a pier, one of the water stations of the Manchester and Lincolnshire Railway, which stretched for some 1,500 feet, like a long black snake, into the Humber as though intended to end only on the other side, though, in fact, the intentions of the engineer architect were more modest. The earliest transit at this spot was by an open boat running chock-a-block upon the beach. Then came a small wooden pier greatly descending to low water mark. Then a tub of a steamer on a narrow, slippery pier ; and now the railroad was extending itself into the stream far enough for passengers on the Lincoln-shire side to make but one step from the steam coach to the steamboat at every state of the tide. A little lower down the stream, fast advancing too, though not so far advanced, a dock of three acres was in progress, intended to be surrounded by sheds for goods and pens for cattle about to be attracted to New Holland ferry by the convenience of transit. Hull is at present the real capital of this part of North Lincolnshire, taking from it a good deal of butcher's meat and supplying groceries and other domestic requirements, young horses, and lean cattle."

The traveller proceeded from New Holland to Great Grimsby, remarking on the various features of the railway and the embankments and the agriculture.

"I made," he says, "my headquarters at Great Grimsby, a port until within the last five years so obscure that it probably owes its place in maps to its privileges as a parliamentary borough."

Formerly the place was of some importance, but it had gone the way of Tyre and Sidon, and our own cinque ports of Hythe, Sandwich, and Romney. Its natural harbour had silted up.

"Within the last few years a succession of events have tended to make the name of Great Grimsby less absurd. The increased use of bones, rape seed, oil-cake, and other tillages largely imported from the north of Europe; the reduction of the timber duties, the alteration in the corn laws, the general reduction of tariff on Baltic produce, and lastly and chiefly, the rapid advance of railway communication, have all tended to revive the decayed fishing village towards a position commensurate to its armorial and genealogical claims. The importation of timber for railway use has quite cast into the shade the profits formerly netted from contraband trade," and generally there will be a great expansion of legitimate industry. "I anticipate," he says, "that warehouses and steam mills for grinding British and foreign wheat, for crushing linseed, and sawing timber, will shortly rise up along the lately dreary quays of Grimsby." In addition to the railway facilities, the directors are "pressing on one of the most complete and extensive docks in the kingdom from the designs of Mr. Rendel, the engineer of Birkenhead Docks."

"After examining these works and a railway map, including not only England, but the opposite shores of the German Ocean, it was impossible to doubt that railway enterprise would make Grimsby a great corn market and a great seaport."

"The fish caught by deep-sea fishing off Grimsby cannot

be surpassed; cod and turbot are to be had in any quantity, worthy of the London market, and may be transmitted by railway to London six hours sooner than if landed in Hull."

Mr. Sidney's prophecies have been better fulfilled than those made by seers of much greater pretensions.

The making of these railways was not without incidents of engineering interest. The system included important works, such as the Wicker viaduct and the Sheffield Victoria Station, the New Holland floating bridge, and two fine bridges over the Trent at Gainsborough and Torksey. With regard to the last a report addressed to the directors of the Manchester, Sheffield, and Lincolnshire Railway by Mr. Fowler on February 15th, 1850, recites that

"the line from Leveston to Saxelby was finished in sufficient time to be opened for public traffic on January 1st, but objections were made by the inspecting officer to the sufficient strength of the wrought-iron bridge across the Trent at Torksey, and the opening to the public has been consequently postponed. The bridge has been tested with a load three times as great as will ever be brought upon it in practice; it is at least as strong as the numerous other wrought-iron bridges upon the Manchester, Sheffield, and Lincolnshire Railway, which had been previously sanctioned by other inspecting officers; and it is of greater strength than the Conway and Britannia bridges on the Chester and Holyhead Railway. Under these circumstances I cannot recommend any alteration to be made to it, and I have no doubt the railway commissioners will shortly rectify the error into which they have unquestionably fallen."

The officer of the Board of Trade who disallowed the Torksey Bridge was the present Field-Marshal Sir Lintorn Simmons, then a captain in the Royal

Engineers. The incident was a serious one for the young engineer, and might have had disastrous consequences for his future career. The matter was argued with considerable warmth, but, as Sir L. Simmons has assured the author, with good temper and fairness. It was the beginning of an acquaintance between the two men, which afterwards ripened into one of the most valued and intimate friendships of Sir John Fowler's later years. The controversy of Torksey Bridge is of some interest and deserves a fuller narration.

Mr. Scott Russell brought the subject before the Institution of Civil Engineers on January 29th, 1850. It was well known, he said, that for some years past there had been many attempts to restrict the free exercise of the talent and ingenuity of engineers, and to interfere with the progress of mechanical and constructive science by the establishment of Government boards and commissions ; almost, in fact, endeavouring to introduce a system analogous to that of the Ingénieurs des Ponts et Chaussées, which has proved so detrimental to all individual enterprise in France. In the year 1847 a Royal Commission was appointed for inquiring into the application of iron in structures exposed to violent concussions and vibration ; it expressly stated, that "considering the great importance of leaving the genius of scientific men unfettered for the development of a subject, as yet so novel and so rapidly progressive, as the construction of railways, we are of opinion that any legislative enactments with respect to the forms and proportions of the iron structures employed therein would be highly inexpedient." Relative to the forms of construction of hollow girders of wrought iron, it was also stated "those methods appear to possess and to promise many advantages; but they are of such recent introduction that no experience has yet been acquired of their powers to resist the various actions of sudden changes of temperature. For the reasons above stated we are unable to express any opinion upon them." Almost simultaneously, however, with the issuing of this report, a girder bridge, built of wrought iron from the designs and under the superintendence of an engineer of admitted skill and extensive practice, was declared by one of the inspecting officers of the Railway Board to be unfit for the public service, because it did not conform to the rules, which, in the report of the commissioners, were expressly declared to be applicable to cast iron only. The actual consequence of this decision, or rather of this application of an antiquated formula to a modern invention, was that the public had already been for one month deprived of the use of an important line of railway, and the probable consequence was the condemnation of the majority of the railway girder bridges, which had for years borne with safety the greatest loads that could be imposed on them under any circumstances of their traffic, and the possible result might be the rejection of that magnificent monument of engineering skill, the Britannia Bridge.

Mr. Scott Russell hoped that the Institution would take steps to expostulate against the wrong under which the public and the profession were suffering. On February 5th he inquired again what the Council was doing. Mr. Vignoles supported his appeal, and complained of the action of the Railway Commission

H

with regard to a bridge built by himself. Mr. Simpson, vice-president, asked members to send all information in their power to the secretary. Mr. Fowler's action in declining to recommend the railway directors to make any alteration in the Torksey Bridge was without doubt supported by the opinion of the leading members of his profession.

On March 12th, 1850, Mr. William Cubitt in the chair, Mr. William Fairbairn read a paper on "Tubular Girder Bridges," with special reference to Torksey Bridge. A difference of opinion, he said, appeared to exist: (1) As to the application of a given formula for computing the strength of wrought-iron tubular bridges; (2) as to the excess of strength that should be given to a tubular iron bridge over the greatest load that can be brought upon it; and (3) as to the effect of impact, and the best mode of testing the strength and proving the security of the bridge. After a technical discussion, in which he criticises some of the proportions of the Torksey Bridge, he sums up :—

"These appear to be the facts of the case, and although the principal girders do not attain the standard of strength which the author has ventured to recommend as the limit of force, they are, nevertheless, sufficiently strong to render the bridge perfectly secure."

The discussion of the paper lasted over several evenings.

Mr. Fowler, who was of course present, expressed his gratification at the result of Mr. Fairbairn's investigation, but pointed out that no allowance had been made for the additional strength of a continuous girder. This, he calculated, would add "one-fourth‧

to the absolute strength of the girder spanning each opening."

In building the first of these girder bridges (the subject being new to him) Mr. Fowler stated that he had been guided by Mr. Fairbairn's proportions, as he was the constructor of the girders, and he complained that Mr. Fairbairn had now changed his views with regard to the requisite proportions, and, as Mr. Fowler maintained, without sufficient reason.

Mr. Bidder, who along with other engineers had also been requested to examine the bridge, reported that after careful inspection and consideration the general opinion arrived at was that the bridge was sufficiently strong for all practical purposes of public safety. That, he gathered, was Mr. Fairbairn's opinion, though he had detracted from the value and weight of that verdict by assigning other proportions for a bridge of such dimensions. He proceeded to question some of Mr. Fairbairn's minor conclusions. Mr. Wild and Mr. Pole had also examined the bridge, and were satisfied that it was of sufficient strength. Indeed, throughout the lengthened debate, opinion in this respect was unanimous, but naturally some difference of opinion existed as to the precise additional value of the continuous girder and other technical points.

Captain Simmons, in the course of the discussion, admitted that he had made no allowance for the fact that the girder was a continuous beam, " because he believed the beam was not so constructed over the central pier as to support all the strain which might be brought upon it, and therefore he thought the paper which had been read, however interesting, did not apply to the case in question."

The controversy dragged on for some months, and month by month the Railway Commissioners postponed the opening of the line. During March, elaborate experiments were carried out, with the assistance of Mr. Wild and Mr. Pole, in the presence of Captain Simmons and Captain Laffan.

"The experiments," says Captain Simmons in his report, "were made with great care, and are therefore to be fully relied on, Mr. Fowler, assisted by Messrs. Wild and Pole, having afforded every possible facility and assistance in rendering them trustworthy."

In conclusion he states :—

"I am induced to recommend that the company be permitted to use this bridge for public traffic, provided the engineer will make such an arrangement of the platform that the ballast cannot be allowed to accumulate beyond the depth of two inches, upon which consideration was based his calculation of the weight of the structure, and also that careful tests should be applied from time to time, with occasional inspections by an officer of this department, who would report whether by the effect of traffic the elasticity of the metal, giving the effect of continuity to the bridge over the two spans, remains unimpaired."

On April 6th the commissioners announced that on these terms they had reconsidered their decision, and formal permission was given for the opening of the line on the 25th of April.

"Thus," concludes the official notice, signed by the Secretary of the Institution of Civil Engineers, "had an important line of railway been arbitrarily closed for a period of upwards of four months, and a bridge been condemned as unsafe which, when examined by practical engineers, had been proved to possess ample strength, and all this in consequence of the attempt to introduce the system of centralisation and of Government supervision, which was found to be so pernicious in continental states, and the employment of officers who possessed undoubted skill for their own military duties, but who were placed in a false position when they were entrusted with the execution and control of civil works, of which their previous pursuits precluded their obtaining a practical knowledge."

Many years afterwards, in an article on Railway Accidents in the *Nineteenth Century* for June, 1877, Fowler alluded to this episode in the following terms :—

"An Englishman is naturally impatient of State interference with any private enterprise in which he has invested his capital, and in the earlier days of railway inspections by the Board of Trade the proceedings of the inspecting officers were suspiciously watched, and sometimes bitterly resented. In one notable case, where considerable public inconvenience resulted from a difference of opinion between the inspector and the company's engineer, official use was made of the circumstance to illustrate the baneful effect 'of the attempt to introduce the system of centralisation and of Government supervision which was found so pernicious in continental states, and of employing officers who possessed undoubted skill for their own peculiar military duties, but who were placed in a false position when they were entrusted with the execution and conduct of civil works, of which their previous pursuits precluded their obtaining a practical knowledge.' Subsequent events have, nevertheless, conclusively shown that State control in the construction of railways does not involve an injurious 'system of centralisation,' but that it is productive of vast benefit to the public, and even to the railway officials, as it provides an independent check of the stability of the works and of the sufficiency and completeness of arrangements for the safe conduct of traffic."

The article from which the above quotation is taken is a very judicially conceived criticism of the report of the Railway Commission of 1874, issued in February, 1877. After making the large admission above quoted, the article goes on to urge, in the terms of the commissioners' report, that "any change which would relieve railway companies from the responsibility which now rests upon them to provide for the safety of their traffic would be undesirable." It dissents, however, strongly from certain recommended invasions of this principle which the commissioners sought to incorporate with it. Their conclusions, Fowler observes, "exhibit verbal dexterity rather than practical wisdom." The difficulty of the commission in arriving at a unanimous decision in respect of these proposed exceptions is pertinently quoted as a reason for leaving responsibility in the hands of the executive.

The concluding paragraphs of the article are worth quoting :—

"It will be gathered from the tenor of our remarks that thorough and searching inquiry into every alleged sin of omission and commission on the part of a railway company, and publicity to the report of the inspecting officers of the Government, are the remedies which we would substitute for the legislative interference recommended by the commissioners.

"If it be said that this, after all, is but a poor guarantee as compared with the legislation deemed necessary by the commissioners, we would say that the influence of what is in our opinion the most important of all of the elements conducive to public safety has been entirely ignored by them, and that the omission invalidates their conclusions. The element of 'human fallibility' has justly received the fullest consideration from the commissioners, but surely it was no less essential for them to recognise the existence of the active

living force 'human sensibility.' How constantly do we hear
of accidents being prevented by the presence of mind, promptness, and energy of railway officials, and of porters and guards
being cut to pieces in the attempt to save passengers from the
consequences of their own imprudence. There appears no
good reason to exclude the directors of railways from the
possession of 'human sensibility,' even as we must admit that
they are liable to 'human fallibility,' and instances might be
adduced without number in proof of the presence and practical
daily value on railways of the element 'human sensibility.'
Yet, in the report, some of the commissioners gravely assume
that the companies actually balance the probable cost of railway collisions against the cost of works that would obviate
them, and that they are guided in their decision by the appearance of the figures! If it is to be taken for granted that
engine-drivers and guards are ready to lay down their lives,
superintendents and managers to endure anxiety and mental
anguish hardly less tolerable than death, and directors to incur
the odium attached to a preventable accident merely that an
eighth per cent. greater dividend may be announced to the
shareholders, then it must be concluded that the purchase of
the lines by Government, or a system of legislation infinitely
more stringent and penal than that proposed by any of the
commissioners, can alone ensure the public against the
occurrence and consequences of frequent and preventable
accidents."

This expression of opinion from a man of Fowler's
practical insight is of considerable importance. We
are in this country irrecoverably committed to free
trade and private enterprise. More and more the
principle of exchange is becoming the pivot on which
the delicate machinery of the fabric of civilisation is
made to revolve. If those whose view Fowler is here
combating are right, the relations of commerce are
absolutely impervious to those generous feelings and

sentiments which, explain them how we may, we recognise instinctively as part of our higher nature. Fowler's protest against this view is timely and weighty. We hear much talk, and reasonably enough, of the good citizen, that is the man who brings public spirit and rectitude to the political and municipal affairs of his country, but the ground covered by our political life is infinitesimal compared with the fields in which effort is organised by the influence of the market. It is a low and, what is more important, a false view of human nature which assumes that an unscrupulous love of gain is a motive predominant in all industrial undertakings. The prejudice against which Fowler is here arguing is a most harmful one. It is unworthy of a great commercial nation like England. Honour and rectitude and fair dealing are qualities evolved not by authority, but by and in the mutuality of exchange. There are dishonest tradesmen just as there are negligent and corrupt officials. There is no escape from the fallibility of human nature and from the occasional predominance of the lower motive, but the rules and restraints of moral sentiment and conscience are as appropriate and, at least, as powerful in the transactions of commerce as in the transactions of the State.

CHAPTER IV.

PROFESSIONAL CHARACTERISTICS AND WORK
1850-1860

WE propose in a later chapter to set out as fully as we can the personal characteristics of the subject of this biography. Following here the course of his professional career, we think it important to notice the somewhat impersonal nature of the work which he helped to accomplish.

The magnitude and multiplicity of a modern engineer's engagements (and Fowler's career practically covers the modern period) made it necessary for him to delegate much of his responsibility, to equip an office, and to organise a staff of assistants. No engineer who did not accommodate himself to this new order of things could be really successful under the modern conditions then beginning, unless he was prepared to kill himself with overwork. The development of Fowler's professional character was naturally influenced by these considerations, and the result we must now endeavour to describe.

In his professional career, the responsibility for great undertakings, which came upon him early in life, obliged him to acquire and to develop a capacity for command.

Apart from his engineering skill, he was emphatically

a good man of business, a character which, by the fashionable sentimentalism of the time, is thought to require some apology. Such a view, however, is superficial. Without the good man of business, neither the accommodation of the public, the wages of subordinate workers, nor the dividends of investors are secure. The master mind that gives order, punctuality, direction, and economy to a number of otherwise incoherent and wasted forces, and which finally crowns their co-operation with a great result, must, to a certain extent, be inexorable and exacting as the laws of gravitation. All these considerations have without doubt their influence in forming the manner and character of men involved in these important responsibilities and employments.

To some extent Fowler was himself conscious of the necessity under which he lay of acting a part. Personal considerations, he would frequently say, ought not to influence our judgment in matters of business. This opinion was based on reason and experience, and was not the natural bent of his character. In his own household he was the most impulsive of men. With his children and with his servants he was indulgent to a fault.

We are familiar with the so-called "Paradox of the Actor," who, when he is depicting the most impassioned frenzy on the stage, is really in his coolest and most calculating mood. There is some parallel to this in what we may call the paradox of the man of business. Here is a man naturally genial and impulsive, consciously or unconsciously making himself an impersonal pivot on which vast industrial operations are made to revolve. It is his function to conceive a great idea, to collect the instruments for carrying it into effect, to lay down the orbits within which each shall move, and to preserve, by a rigid enforcement of the punctual and due performance of separate contracts, the connected harmony of the whole. The man who is accustomed to such great operations, who sees how their successful conduct leads to the happiness of mankind, soon learns the necessity of banishing whims and personal caprice from his moods, and naturally acquires a certain magnanimity of view.

In discussing the career of his friend and partner, Sir B. Baker has summed up to the author the nature of Fowler's success, in the statement that very early in his professional career Fowler somehow had acquired the position of being the man to whom persons with a big project on hand felt obliged to apply. We have it on high authority that there is no new thing under the sun, and the truth is one which would be confirmed by a study of the history of great engineering works. The scientific imagination is concerned not so much with new things as with new applications of well-known principles and facts. Further, the scientific imagination can achieve very little till it is materialised by an organising faculty which is not imaginative, but full of the sober prose belonging to the management of men and finance.

Fowler was by no means devoid of the scientific imagination, but undoubtedly the aspect of his character which impressed the public was his strength and dexterity in carrying great proposals to a successful issue. It has been pointed out that a great invention or a great idea may lie idle till it is connected with some motive force which thrusts it into use, in spite of the initial frictions which stand in the way of all improvement and change. This truth contains the justification of patent laws when regarded from the point of view of the general public. It is an instance of the way in which the security of private interest is rendered subservient to the public welfare. The peculiar position attained by Fowler seems a further illustration of the same truth. The idea of tunnelling a railway through London is obviously the result of many precedent suggestions, but it lay dormant for long, in ingenious but ineffectual hands, and only became actual when the motive forces to carry it into execution were organised and set in motion under the guidance of Mr. Fowler.

As a man of affairs Fowler had many remarkable characteristics. As we turn over the minutes of evidence submitted by him to parliamentary committees, it is impossible not to be struck by his skilful presentment of a case; his replies to examination in chief are clear, concise, and to the point, but it was in cross-examination that he was most persuasive. There he was deferential and appreciative of the points that were raised against him. The objection raised, he would say, had been duly considered by him, indeed, it was chiefly because his scheme overcame it so completely that he spoke so strongly in its support; or if this line of reply was not applicable, he would urge that there were difficulties in every alternative, and that although the objections were put with great force, they were more easily superable than those which beset other schemes. There is no attempt at display, no desire to browbeat or to be witty at the expense of his questioner, nothing apparent but an eager desire to bring out the truth by means of fair discussion.

A director of one of the railway companies with which Fowler was connected bore the following testimony to his persuasive powers: "I never met any one," he said, "who so often was able to convince me that things which I had all my life long regarded as white were really black." There was a genial sympathy about him that was irresistible. In person, Fowler was a big, powerful man; his habit of speech was direct, yet so kindly and humorous that it never suggested aggression. Nor did his appearance belie his character; he was emphatically a strong man, and the strong man who can also be conciliatory as a rule will go far.

It has been remarked by one who knew and admired him that he seemed at times to attach an undue importance to material success. The remark raises a fundamental problem of ethics. If, as the evolutionary school of thinkers maintain, ethical sentiment is the approval of a body of rules and motives dictated and sanctioned by the experience of human society, a man with the experience of Fowler must have observed that industry, punctuality, the honourable performance of contract, the considerate and sympathetic rather than the coercive management of men, self-control and self-renunciation, the power to resist the temptation of snatching illegitimate personal advantage, are the qualities which lead to reputation and permanent success. In regarding success, therefore, as in some measure an index of worth, Fowler did not necessarily take a low view of life.

The point is of some importance, for Fowler seems to us to be a typical figure. With many personal and unaccountable idiosyncrasies thrown in, he was a product of the time. His success came to him because he was fully in accord with the spirit of the new industrial era. It is a vindication of the Economic Order that a man obviously drawing his inspiration from that source is found honourable, veracious, and courageous, in his personal relations kindly and courteous, and, what perhaps is equally remarkable, dignified in his leisure by a large and enlightened appreciation of all that is best in the social life of the time.

There is no trace in his letters and recorded conversation of any introspective questioning of his own motives and maxims of life. They seem to have been more or less the spontaneous outcome of his own experience. His native conservatism and his constitutional respect for that which is and must be, his aversion from revolutionary views, made him a strong supporter of Church and State, but he was not a man whose deliberate actions were dictated by effusive sentiment, nor was he given to that abstraction of thought which is characteristic of martyrs and reformers.

He was, in short, an example of a man of the strictest integrity with whom the sanctions of right conduct are practical intuitions rather than a rule of life based on religious or philosophic theory.

The following appreciation of his characteristics as an engineer is compiled from information given by those who were for many years associated with him in business.

He was professionally a man of great moral courage, very prompt to face a difficulty. It is a very usual thing for engineers to follow precedent closely in

difficult situations. Fowler, however, was very courageous in making new departures. Thus, notwithstanding his admiration and friendship for Brunel,
when on the death of that great man he was
appointed his successor as consulting engineer to the
Great Western Railway, he urged the directors to face
the loss of entirely abandoning the broad gauge.

He took the lead, also, in recommending the use of
iron or steel in bridges instead of timber, and also
he was among the first to urge the employment of
steel instead of iron rails. His use of Portland cement
concrete for retaining walls was also an innovation.

A well-qualified critic thus sums up his professional
character :—

"As an engineer he had of course many equals from a
scientific point of view, but he was so eminently practical
that nothing beat him, and his power of dealing with men
was wonderful. He had the faculty of 'spotting' a fault
in an opposition scheme better than most men, and this
applies also to drawings or designs brought to him by his
own assistants for approval."

He used laughingly to say that if he did not
understand a thing at once he never understood it,
and certainly his rapidity of judgment was very
remarkable.

It is no part of our plan to gibbet the subject of
this biography as that abnormal monster a perfect
man. In the course of his professional career Sir
John Fowler had a good many controversies, which,
as might be expected, he pursued with much determination ; for though there was nothing vindictive in
his nature he was very unbending. He had, indeed,
a considerable share of that quality which the in-

dulgence of our friends calls firmness, but which in hostile circles is most unreasonably regarded as obstinacy. This quality of heart was with him so spontaneous that he was by no means conscious of its existence. One of our informants tells us that in confidential talk he expressed the most unbounded surprise in detecting what he considered a trait of obstinacy in one of his children, a quality, he remarked, so entirely absent from the character of his parent.

Of his dealings with contractors, one of the most difficult personal tasks of the engineer, somewhat various accounts have been given. With the big and powerful firms, such was the keenness of his lust for battle, he was more inclined to be exacting than with smaller men. In the initiation of engineering projects, the contractor is a very important element. It is often he who furnishes the funds for the preliminary expenses, and at one period of his career Fowler suffered a good deal from his inability to agree with certain large and prominent firms. With foemen worthy of his steel he was more inclined to thresh out his differences than to accommodate matters by compromise or concession. The precise equity of these controversies is not now of any importance. Such as they were, however, they were undertaken rather as the result of temperament than from any desire to snatch an illegitimate advantage. With the smaller men, we are informed by one whose relations with Sir John were not always smooth, he was fair and, as far as his clients' interest would allow, liberal.

The same characteristics were apparent in his relations with members of his profession and with his subordinates. In matters where the responsibility was entirely his own he naturally would brook no interference. In the many relations of an engineer's business where the responsibility is more or less divided, he was ready, perhaps eager, to assume a dictatorial position, a course not to be wondered at in one of his self-confidence and long experience of success. He was probably right in thinking that, if his supreme authority was conceded, the matter in hand would be better managed than under a divided direction. On the other hand, if his authority was questioned in departments where the responsibility of one of his colleagues was clear, he acquiesced quite good humouredly in the situation, and recognised that there might be other masterful men besides himself. He recognised, in fact, that there were differences in men, and accommodated himself to that which he could not bend. It must, however, be admitted that he liked to be the predominant partner. Being himself painstaking and industrious he expected his subordinates to be the same. No one understood better than he the need of delegating work and responsibility. His habit was to select his assistants with care, to put them in charge of considerable undertakings with a minimum of supervision and direction, and to trust them. To have been one of Fowler's assistants was a certificate of proficiency, and a large number of men who have done and will do their country good service have been associated with Sir John Fowler as his trusted assistants. In this way we find that he had expeditions to France, Spain, Portugal, Algiers, Italy, and over a series of years to Egypt, the Soudan, and the Upper Nile.

I

There are two sides to most questions, and the
relation between a leading engineer and his assistants
is no exception. In the devolution of work to subordinates a slight miscalculation may easily throw too
great a burden on individuals, and the demand of the
organiser may appear exacting. The good organiser,
however, will in most cases fit the task to the workman,
and Fowler's skill in this respect was certainly extraordinary. Not a few of those employed by him in
special investigations have become recognised authorities
in those particular lines of work. This selection of
instruments is of course a part of the patronage at the
disposal of a successful engineer, and in a profession
abounding with talent and energy the selected subordinate is to a certain extent put under an obligation
to his chief. He is given his opportunity. On the
other hand, the period of apprenticeship and sub-
ordination has a way of seeming unduly long to the
ambitious young aspirant and to his friends ; but, as
Adam Smith long ago pointed out, in every free
bargain both sides profit, and there is after all no more
equitable method of settling the value of services than
the give and take of the market.

Fowler was very fortunate in the selection of his
business associates. With many of them his connection
extended over a very long period. His distinguished
partner, Sir Benjamin Baker, entered Mr. Fowler's
office in 1861, and became a member of the firm in
1875, and the record of their co-operation fills a large
chapter in the history of English engineering. Notwithstanding his undoubtedly imperious character he
had the power of attaching men to him by very warm
ties of admiration and affection.

"Sir John's disposition," says one who knew him well, both in business and at home, "was a very affectionate one, and his home life was very happy. He was a loyal friend, and I do not remember to have ever heard him speak ill of any one. He did not by any means make a friend of every one, but he did not speak against those who were not his friends, or against those who had offended him."

There is a complete absence of anything in the nature of personal gossip in his letters, and he had, we are informed, a very strong objection to what he called talking people over.

The following letter to his old friend, secretary and partner, Mr. Baldry, might appear to be an idealised reminiscence rather than a plain statement of fact. Mr. Baldry, however, assures the writer that its terms do not appear to him exaggerated.* Such a letter throws a pleasing light on the prosaic routine of the office at 2, Queen Square Place :—

"THORNWOOD LODGE,
"CAMPDEN HILL, KENSINGTON, W.,
"*November 8th*, 1888.

"MY DEAR BALDRY,—I wished to say a few words of good-bye and good wishes to you last night in your own office before our final business separation, but I saw it was too much for you to bear, and I must now ask you to let me write the words which your emotion did not permit you to hear.

"It is, I am sure, permitted to a very limited number of men to have had the intimate and confidential business relations which have for thirty-six years subsisted between you and me, and for both of us to be able to say at the end

* While these pages are passing through the press, the death of Mr. Baldry at Hyères is announced (Feb., 1900). Mr. Baldry's amiable and kindly nature endeared him not only to his partners, but to all who came in contact with him.

of the time that not one cool look or one unpleasant word
has ever been known during the long intercourse.

"But the word business expresses very inadequately our
relations. During the last twenty-five years at least you
have really had charge, not only of the private affairs but
almost the consciences of every member of the family, and
have frequently rendered, as we could all testify, the most
essential service.

"Lady Fowler feels everything I would say, but which
it is impossible fully to express. I know every son, and
indeed every member of the family have the same feelings
of regard and gratitude as ourselves.

"I can speak for all the staff in the office, and indeed for
all Westminster, and the numerous persons who come to Queen
Square Place, when I say that no man ever left the scene of
a long and honourable career with more unanimous wishes for
future happiness than you carry away with you.

"But although, my dear Baldry, your health and prudence
require that you should now leave your business chair at
Queen Square Place, you know that my friendship and that
of my family is for life, and it will always be the great hope
and pleasure of Lady Fowler and myself that you should
frequently be our guest both in London and at Braemore.

"I could write many pages, and then I should not tell you
all I think and feel; but you will know our regard and
affection for you, and how sincerely we wish you health,
happiness, and every good wish in your future years.

"Believe me, my dear Baldry,

"Yours very truly,

"JOHN FOWLER."

This letter is somewhat of an anticipation, but it
is quoted here to show that though a man of Fowler's
combative zeal and ubiquitous energy did not
altogether avoid the giving of hard knocks and the
incurring of enmities, he was personally a man of a
very lovable and affectionate disposition.

In the year 1850 Mr. Fowler married Miss Elizabeth Broadbent, daughter of Mr. James Broadbent, of Manchester, a lady with whom he had become acquainted the year before. In many respects John Fowler is to be described as a fortunate man, but in nothing was he more fortunate than in his marriage. The copious correspondence which has been placed at the biographer's disposal covers a period of nearly fifty years, and constitutes a continuous testimony to the warm affection and entire confidence which characterised the relations between husband and wife. Mr. Fowler, owing to the calls of his profession, was frequently absent from home. During these absences he worked with fiery energy, rising at daybreak and putting an almost incredible amount of work into every twelve hours, but amid all this pressure of business he never omitted his daily letter to his wife. Though for biographical purposes they contain little that is of public interest they are charming letters, and throw a most agreeable light on his domestic life and on the kindly, considerate, and courteous character of the writer.

Their first home in London was 2, Queen Square Place, a roomy, old-fashioned house adjoining Bird Cage Walk. Mr. Fowler's offices were under the same roof as the dwelling-house. There was a large garden, and before the building of Queen Anne's Mansions the situation was cheerful and bright. The house had a historical interest. Milton had lived hard by, and had planted in the garden a cotton willow tree, which was blown down during the Fowlers' tenancy. The wood was made into a "davenport," and is still a valued ornament of Lady Fowler's drawing-room. The house itself had been the home of Jeremy Bentham, and the meeting-place of the two Mills, Chadwick, Brougham, Place, and the other founders of the party of philosophic radicalism. It was in this garden that the old philosopher used to take his exercise, ambling round the walks generally in the company of one of his disciples, and, as he put it in his pedantic way, "maximising recreation and minimising time."

Here the first years of Mr. and Mrs. Fowler's married life were passed, and the place and the actual address had a sentimental value in his eyes, which may appear strange, but which, nevertheless, is entirely in keeping with his character. Later the ancient landmarks of the immediate neighbourhood were removed, and the enormous block of Queen Anne's Mansions was built upon the site. Sir John Fowler retained in the reconstructed buildings an office which is still known as 2, Queen Square Place. There is no number one, nor indeed any other house in the "Place." There are other and better known Queen Squares in London, and the address occasionally was found misleading by hurried clients, who arrived at times in a ruffled state of temper which did not facilitate business. His partner, Sir B. Baker, tried hard to bring about a change of name, and at length one evening obtained his consent, but next morning Fowler returned, and in a most pathetic manner asked his younger partner to yield to what he called the foible of an old man, and 2, Queen Square Place the office remains to this day. The addition of "Queen Anne's Mansions" to the firm's address was accepted as a compromise.

To return to the year 1850. In one of the earliest letters to his wife in the first year of their marriage he

gives a passing and not an unpleasing glimpse of a
notoriety who was much before the public at that time.

"I travelled alone," Mr. Fowler writes, "with the great
Mr. Hudson from London to Normanton and had a great deal
of conversation as to his troubles.

"He told me the greatest comfort he had when all the
world seemed against him was his wife, and he thought that
if he had not received that comfort it would have been too
much for him. I told him I was a very new married man,
but I could entirely feel the force of his observations. . . .
And now, my dear little wife, adieu, and may God bless you
with all the happiness your husband wishes you. I feel
already how essential you are to my happiness, and I am sure
before Thursday I shall be very impatient to be back again."

The career of Mr. Hudson, the Railway King, ran
altogether apart from Mr. Fowler's work, but it is
characteristic of Fowler's good-natured tolerance for
a man who was both courted and condemned beyond
his merit, that in his numerous passages to the
Continent he used from time to time to seek out the
ex-Railway King in his dingy retirement at Boulogne
and try to enliven his exile by a little friendly hospitality. Mrs. Hudson, a blameless and exemplary lady,
lived to old age on Campden Hill, Kensington. Sir
John and Lady Fowler, then living at Thornwood Lodge,
and not forgetful perhaps of this little incident, were
glad to avail themselves of the privilege of neighbours,
and by various kindly attentions to show their
sympathy for misfortune borne with dignity and
resignation.

Of engineering matters his letters to his young bride
naturally say very little. A letter of October 29th,
1850, records a visit to New Holland, and mentions
"a most successful trial of our new crane." These

were hydraulic cranes, designed by Sir W. Armstrong, and, we understand, were among the first mechanical appliances of the kind used for the handling of goods on railways, and we should have been glad to hear more of them.

In January, 1851, he paid a visit to France, in which Mrs. Fowler accompanied him, with regard to the Douai and Rheims Railway, a project which came to nothing. On December 2nd, 1851, he records the *coup d'état* of Louis Napoleon, with the comment, "strong measures, but I believe necessary."

On June 1st, 1852, he writes to tell his wife that overtures have been made to him by the people of Dudley with a view to his representing them in Parliament.

"I find a curious feeling at Dudley on the subject of the election. I speak quite seriously when I tell you a very influential party are anxious to get up a memorial to ask me to be a candidate at the next election, if there is the slightest prospect of my receiving it favourably, and certain parties in the district who are friends of mine are to have a non-official interview with me in a few days for the purpose. I understand from various circumstances I should stand an excellent chance of being returned, and free of expense. Of course I should not do anything without consulting you. What say you?"

Mrs. Fowler, who was better aware perhaps than her husband himself of the great strain which the pressure of his engagements made on his strength, was strongly opposed to his adding parliamentary to his professional work. A few paragraphs from the correspondence between the husband and wife will show how the matter was allowed to drop.

"I never, never could consent willingly," says Mrs. Fowler, "to your undertaking a seat in Parliament. I should be very unhappy if you were to do so. Do not be sanguine about it, my darling, but rather throw cold water on it in your mind and give it up at once.

"I cannot see that the advantage which it might be to you is worth the consideration. Position you have, and you do not want more. Time you have *not* to devote to it, and instead of involving yourself in larger and greater engagements, let them diminish and enjoy yourself more. Take more leisure and pleasure, rather than confine yourself more by undertaking such a responsibility. More I will not say till I see you, which I hope will be to-morrow, when you will indeed have a welcome."

In deference to his wife's wishes, Mr. Fowler took no steps at this time to enter Parliament. Some years later, in 1859, the project was raised again, and Mrs. Fowler a second time urged the unwisdom of such a step. His reply to her is as follows :—

"I appreciated all your letter very much excepting that part which tells me you did not think you had the same influence over me you formerly had. This is a great mistake, and to prove it to be so I will undertake not to take any steps to go into Parliament except with your approval.

"Keep this letter, and bring this forward against me if I am disposed to take any steps against your wishes."

He seems about this period to have turned his back on all thought of a parliamentary career, and, in accordance with Mrs. Fowler's advice, to have taken the first steps towards acquiring that Highland home which was for so many years an abundant source of pleasure to himself and his friends.

In 1857 he purchased the estate of Glen Mazeran, near Tomatin, in Inverness-shire. On the 17th of September of that year he writes to compliment his father on a correct estimate of the letting value of one of the farms on the estate, and describes his first experiences as a Highland laird :—

"Yesterday I went out deer stalking on the upper part of Glen Mazeran, leaving the lodge at six o'clock as the sun was rising over the mountain, but as I rode up the glen the wood was in shadow and was quite hard with frost, and the water on the sides of the road was frozen nearly a quarter of an inch thick. In some places on the higher ground, even for an hour after the sun had risen, the ground was quite crusted over by the frost, and cracked and split when trodden upon as in winter.

"At nine or ten o'clock, however, the sun became sufficiently powerful to warm the earth and its inhabitants.

"I must tell you, for the information of Fred, that I found two splendid stags, one with ten points which I could easily have stalked, but they were a short distance over the march and my conscientiousness prevailed and I refrained, notwithstanding Fraser's deep disgust and strong remonstrances.

"I was rather sorry afterwards that I had been so very good, for some people we had at lunch told me it was not the habit to be very strict about marches with deer, and with regard to the place where I saw the deer (near Cairn Gregor) nobody was shooting the ground this year, and it belonged to a trustee who is now in America. Glen Mazeran now has always deer in it, since the sheep have been taken off, and I daresay I shall get another stag before I leave, and if I do I will write Fred a description. I wish he was with me now. I want a companion who is not afraid of early rising, walking, and working, when sport is to be had."

His professional engagements from 1850–1860 continued to hurry him hither and thither throughout England, and occasionally to the Continent. A few only of the most important of these can be mentioned. The construction of the Oxford, Worcester, and Wolverhampton section of the Great Western Railway, the London, Tilbury, and Southend·Railway, in conjunction with Mr. Bidder, the Severn Valley Railway, the Much Wenlock and Coalbrookdale Railways and the Craven Arms Railway, a system which included two fine cast iron bridges over the Severn, the Edgeware, Highgate, and London Railway, the Alexandra Park extension, and the Barnet extension, the Mid-Kent route, and the Hammersmith and City Railway, from Paddington to Hammersmith, belong to this period. This work was varied with visits to Paris in October and to Lisbon and Cintra in November of 1855. In May, 1856, he visited Dunkirk in company with Mr. Brunel. In July he delivered a report on the Chester Holyhead route. In July and in December he was in Paris in connection with the Algerian railways. In October he reported on the Turkish railways, and in the same and subsequent years he was much occupied with the Nene Improvement Commission, and the reclamation projects of the Norfolk Estuary proprietors.

The following extract from a letter to his wife gives an idea of the rate and pressure at which he worked. It is dated from Tavistock, March 31st, 1861 :—

"As I travelled to Exeter on Friday afternoon I tried to work out a plan by which I could arrive at home by a train which reaches Paddington to-morrow afternoon at six o'clock, so that I might dine and spend the evening with you; and I found by making a start at daylight yesterday morning and working industriously till late last night, and again starting at daylight to-morrow morning I can accomplish it. So you may expect me at Paddington at six o'clock, and I think you had better dine at seven, and invite Mr. Johnson and Mr. Baldry to dine with us, and then I can comfortably transact all necessary business with them and enable me to go to Ireland without feeling anything neglected."

From 1856 onwards he was closely connected with the development of Irish railways. He was the engineer of the Great Northern and Western (of Ireland) Railway, and visited Ireland yearly in connection with the business of that and other lines of railway. A letter of September 9th, 1856, describes his method of travel. The party purchased a carriage, and attempted to travel in comfort. There were difficulties at each stage about horses, and the harness *more hibernico* was, for the most part, a patchwork of rope, but the weather was fine and the journey enjoyable. The accommodation was primitive and often uncomfortable. He notes that their food—mutton, pork, roast goose and pudding—was all put on the table at once, and in none too cleanly a condition, and that the solid wings of the goose found their way into his feather bed. In later years the proverbial hospitality of the Irish made his journeys more comfortable, and he describes it as "lionising through Ireland, staying at the best houses." It is indeed characteristic of the man that comfort and discomfort interfered very little with his thorough enjoyment of life and work.

In this account of his first journey he describes their arrival at the west coast, where, between Westport and Castlebar, they met the chairman of the line, Lord Lucan. Of the perennial Irish problem Fowler says very little. Irish poverty is very poetical and picturesque, and civilisation is hard and orderly, and not readily adopted by the "finest peasantry in the world," or indeed by any other proletariate class. To Fowler the problem seemed simple enough, and his remedy was briefly : " Let improving landlords introduce better methods of cultivation, we must face the suffering which this change will make. This will be kindest in the end." The economy of western civilisation seemed indeed to him to be as axiomatic as the rules of arithmetic. No plea of Celtic nationality had in his eyes any relevance in abrogating the authority of either the one or the other.

"Between Castlebar and Westport," he says, "we saw the effect of a good deal of Lord Lucan's proceedings. Whole villages completely removed and not a vestige remaining, and fine fields with rich crops of grain and grass growing in their place. It only needs a ride through this country to satisfy anybody this is the true policy, and in the end, kindness to the individual. Only conceive a miserable cabin, ten times more so than the worst you ever saw in Scotland, and ten times more filthy, inhabited by pigs and poultry, also with a swarm of children, and the whole family supported out of a rood, perhaps, of land. Why, of course, fever is hardly ever absent, and starvation never, and the poor creatures are useless to themselves and everybody else."

We need not dwell at length on his work in Ireland, which continued for many years, unless it is to record an event to which he himself attached considerable importance—the capture of his first salmon.

"Yesterday morning I came down to Galway at seven o'clock with Mr. Roberts (who was on his way to Athenry) to try for a salmon on the river, and was fortunate enough to hook the largest fish which had been caught for a fortnight, and after fifty-five minutes of hard work I brought him safely to land.

"I can quite understand now why people become so excited about salmon fishing, and to see a 16 lb. salmon, when it is first brought out of the water, with colour exactly like humming-birds, and to feel you have conquered him, and done it with a single gut—as the fisherman calls it—is like every other success, very pleasant.

"There was quite an excitement on the river, and I was pronounced a skilful fisherman."

His knowledge of Ireland and his well-established reputation brought him, in 1867, the following flattering proposal from the Government :—

"MY DEAR SIR" (wrote to him the Chancellor of the Exchequer, Mr. Disraeli, on August 17th of that year)— "H.M.'s Government are anxious on the subject of Irish railways. They have resolved to issue a Treasury Commission to inquire completely into this subject. My opinion is that it should be limited, and even very limited in number. I should prefer *three* to a greater number of members : a first-rate civil engineer, whose name would inspire confidence ; a first-rate official financier ; and, if that be possible, a first-rate railway manager.

"It will be a paid commission.

"I should be very happy if I could induce you to be a commissioner, and I think it would be in my power to associate with you a first-rate financier. I am quite at a loss with regard to the character who should complete the triumvirate. No doubt there are many first-rate managers if abilities only were requisite for the duties I contemplate.

"Perhaps you might assist me with a suggestion ?

"The main duties of the commission will be to ascertain the financial position of the Irish railways, to accomplish an official verification of their accounts, the state of their lines, their assets.

"I repeat my great desire that I may induce you to assist the Government by undertaking this inquiry. No expense shall be spared in securing for you all the subordinate aid you may require. Let me hear from you at your convenience, and believe me,

"Faithfully yours,

"B. DISRAELI."

To this Mr. Fowler replied :—

"I cannot but feel flattered by the selection of my name and the manner in which you introduce the question to me, and although I am much engaged with my professional duties, I feel it proper to say that if you think I can be useful I will undertake the duties you ask me to perform, but I am afraid I could not make a commencement during the month of September.

"I think the constitution of the commission you propose is more likely to work well than a much more numerous one would be, but no doubt the chief difficulty will be the selection of a railway manager.

"Of the men actively engaged upon the large railways, I think Mr. Seymour Clarke, the general manager of the Great Northern Railway, would probably be the best, both as respects general capacity, public confidence, and other qualities, but it is possible you may feel a difficulty in asking a gentleman who is attached to any particular system of railway, and his company might possibly have a difficulty in sparing him. On these points, however, Lord Colville, one of the directors of the Great Northern Railway, might be safely and confidentially spoken to.

"Of the chairmen or other directors of railways having practical knowledge of the working of lines, but not being actual general managers, I am at a loss to make even one confident suggestion, for so many of the prominent men have lately brought themselves and their companies into trouble by clever 'financing' that very few men are really left.

"In many respects, however, I think Mr. Moon, the chairman of the London and North Western Railway, would be a good selection. He is thoroughly acquainted with railway management and its minutest details, and on the largest scale, and the London and North Western Company has, without doubt, been successfully established under his rule as the best and safest railway property in England."

In the sequel, the gentlemen appointed by a Treasury Minute, dated October 15th, 1867, to serve on this commission were : Sir Alexander T. Spearman, Bart., John Mulholland, Esq., John Fowler, Esq., c.e., Seymour Clarke, Esq., Christopher Johnstone, and W. Neilson Hancock, Esq., ll.d., Secretary.

The possibility of promoting the development of Irish industry by different forms of State aid was then, as now, engaging the earnest attention of statesmen. The task imposed on Mr. Fowler and his colleagues was not to pronounce an opinion on the economic wisdom or otherwise of State ownership of railways, but to collect and arrange certain details of information, and *inter alia* to calculate the amount of loss which would be sustained by the State, if it purchased the Irish railways with a view of reducing the charges to such a low level that an impoverished people like the Irish should become as frequent travellers and users of railways as the Belgians.

The reply of the commissioners to this very hypothetical conundrum is couched in very cautious terms. The total share capital and borrowed money invested in Irish railways was stated to be £27,527,286, and for canals a sum of £902,918 must be added. In their second report the commissioners estimate that the annual loss to the Government if, after purchasing the railways, it reduced the Irish rates to the Belgian level, would be £655,265, or about 42 per cent. of their receipts from passenger, goods, and live stock traffic. The concluding paragraphs of the second report point out that the circumstances of Belgium and Ireland are not analogous. In Belgium the traffic is chiefly connected with mining and ironworks. In Ireland it is almost wholly of an agricultural character, and the movements of a people engaged in that industry àre much more limited. The reduction of charges had been a great success in Belgium as regards long-distance traffic. The traffic of Ireland, however, "requires special stimulus and development for short and moderate distances, both for goods and passengers, as, except in the case of tourists and people travelling for pleasure, it is entirely of local character from town to town." The commissioners therefore think that any reduction, to be beneficial, must be on "long" and "short" traffic equally.

They had no difficulty in coming to the conclusion that a slight diminution in the charges for goods and passengers would be simply a loss of so much money. A large reduction was necessary to create an increase of traffic. In passengers' fares they suggest a reduction of 31, 45, and 42 per cent. for first, second, and third class passenger fares respectively; a reduction of from 47 to 78 per cent. for different classes of goods, and of 32 per cent. for cattle traffic. The deficiency per annum resulting from such a revolution of charges would be £525,701, or 30·45 per cent., after allowing £120,000 per annum as the estimated reduction of expenditure in respect of interest and cost of management.

In the judgment of the commissioners, after the lapse of eleven years the receipts from the increased traffic, called into being by this reduction of fares, would suffice to pay working charges, interest on borrowed money, and on capital advanced to meet losses incurred during the eleven years of loss, and would leave a balance in favour of the Exchequer.

If, as is frequently said, the rise and fall of railway traffic is the best index of the country's industrial

K

progress, this must seem a very simple way of bringing material prosperity to Ireland. The commissioners, however, do not commit themselves unreservedly to this proposition, and add :—

"We do not feel it to be within the spirit of the instructions which we have received to speculate upon the degree of material prosperity which would be given to Ireland by the adoption of a great reduction of rates and charges, and a concentration of management."

Arguing, however, from precedents which are, as the commission points out, totally dissimilar, an increase of traffic may be expected.

"These," the report concludes, "are vast results, and it must be always remembered that calculations of this nature are subject to disturbance from unusual or unexpected circumstances, but having obtained the best and most accurate information in our power, and having brought our own experience to bear upon the questions submitted to us, we do not hesitate in giving our opinion that such results may be fairly expected to follow the suggested reductions."

Everything connected with the economic history of Ireland is deeply interesting. The Government, as we know, took no action in the direction indicated by their desire for information on this subject. The industrial condition of Ireland, if not "unexpected," is certainly "unusual." A mere reduction of railway charges to a population which is not permeated by the industrial instinct, the Government probably judged to be a vain expedient.

To sum up the verdict of the commission, the lowering of rate indicated would, according to precedents gathered in industrial communities, produce the stated increase of traffic in the stated time. This was the reply to the problem proposed to them. Incidentally they hint that possibly Ireland is not an industrial community, and that therefore their calculations may be irrelevant.

The question of Irish railways continued to occupy the attention of Government. In the beginning of 1872 we find that Mr. Fowler was engaged in negotiations with certain eminent financiers for an amalgamation of Irish railways. The conditions on which, in his judgment, negotiations might proceed are summarised as follows in a memorandum drawn up by him about this time : All Irish railways to be purchased ; the Government to find four-fifths of the purchase money at $3\frac{1}{2}$ per cent. ; the proposed company the remainder. The company to provide at least £5,000,000 of unguaranteed capital ; a reduction of 10 per cent. in railway rates ; all rolling stock to be made and repaired in Ireland ; Irish representatives to be on the Board of the company.

The Government, he states, was very anxious about the matter, and would, he believed, be propitious to any suggested plan which should include *Government aid without financial risk or the necessity of Government management.*

"The Irish railway problem" (so the impossibility of a cheap railway system among a population whose industrial capacities are undeveloped is euphemistically termed) was not to be solved in this way. It is of course impossible for a Government to buy up a struggling industry and to reduce its tariff without incurring both risk and responsibility.

It is worth noticing that many years earlier, viz. in

1847, Lord George Bentinck had made a somewhat similar proposal. In what his biographer has termed the best speech he ever made, he urged that the Government should advance £16,000,000, to which £8,000,000 of share capital should be added, for the extension of railway enterprise in Ireland. An interesting account of Lord George Bentinck's proposal will be found in Lord Beaconsfield's life of that statesman. Its revival, when the biographer had become the leader of the House of Commons and Chancellor of the Exchequer, is probably more than a coincidence.

This anticipation of events is, we hope, justified with a view of bringing into one continuous record Sir John's connection with Ireland. To return to the earlier period, we find in 1857 the first entries referring to the Victoria Station and Pimlico Railway Company, as it was called, a scheme which was to make Sir John Fowler responsible for one of the most important exits from London, namely, the Pimlico Railway Bridge and Victoria Station. This bridge was the first railway bridge over the Thames in London, and (we are fortunately able to append an illustration) it is still the handsomest. This encomium, the cynic will say, does not convey high praise; but a review of the London railway bridges will make us wish that Sir John Fowler had been responsible for more of them.

One other expedition of considerable importance was undertaken by him at this time. In 1857 he made a prolonged stay in Algiers with the purpose of constructing a system of railways there. He had interviews with the Emperor Napoleon and his ministers, and made great progress in his survey. The work

THE VICTORIA BRIDGE, PIMLICO

The first Railway Bridge across the Thames in London, opened for traffic in 1860

To face page 132

was finally, by arrangement, given over to French engineers, and the line was not constructed by Fowler and his English assistants. During his absence he wrote long and detailed letters, descriptive of the country and scenery, to his wife. The letters are remarkable letters, both in respect of the historical information they contain and the description and shrewd comment which they make on the country and the people; but they are reticent as to business, and though they suggest to us that Sir John Fowler, if he had given himself time, might have written admirable volumes of history and travel, they are only the record of an uneventful business journey, and contain nothing that is of interest to the general reader.

CHAPTER V.

FOWLER'S work, however, was not confined to railway construction. Hydraulic engineering, from his early training, always had a special fascination for him.

His connection with the Nene Improvement and the Norfolk Estuary reclamation scheme brought him into relation with what may well be termed a perennial and classical problem of English engineering.

The history of the drainage of the Fens dates from Roman times, but its more modern phase began when Charles I. employed Vermuyden on engineering work in this region. Since then the restless and ever varying action of the waters and the sediment which they contain have taxed, generation after generation, the skill and patience of a succession of great English engineers. Sir John Rennie, in a speech delivered on the subject at the Institution, enumerates the great names of those who had given attention to the problem — Vermuyden, Westerdyke, Kinderley, Armstrong, Labelye, Golborne, Watte, Elstob, Smeaton, Mylne, Page, Huddart, the elder Rennie, Robert Stephenson, and, in due succession of time, and on a subdivision of the great question, John Fowler.

The general nature of the problem is well stated by Walker and Craddock in their *History of Wisbech and the Fens*, 1849.

" The Fens are a large basin which, but for drainage, would be generally a standing pool formed by the washings of the high borderlands. The great reservoir into which they discharge their contents is filled with sediment, which every wind and tide throws into the mouths of its rivers. If the Fens were of greater elevation, the force which these elevations would give the fresh waters falling into their outfall would effectually cleanse whatever the tide and tempests might deposit, and—as in the Rhone, which brings vast quantities of earthy materials from its mountains—drive the suspended matter forward till it subsided in deep water. But the Fen rivers are placed in such peculiar relation to the sea—their force is so nearly balanced by the sea force—that they have no energy to push forward, but become passive and clogged up unless art brings in its aid to assist them.

" In these circumstances, whatever force can be imparted to the Fen waters requires to be economised and used as effectually as possible. It is clear that the rivers which have the greatest quantity of fresh waters to scour them have the greatest chance of keeping their channels open, provided these channels are restricted against weakening themselves with breadth. It is, then, an exceeding waste of impoverished means to send the fresh waters of these regions to sea in three or four separate streams, instead of combining them into one stream and driving them into their receptacle in a firm, united body. That this was the original method of nature seems pretty evident from the history of the Great Ouse, which . . . discharged itself originally by Wisbech, carrying with it the waters of the Nene, and even part of those now discharged by the Welland. Engineers have not been entirely blind to this fact, but the extent of the work and the adverse interests it would involve have made them shrink from any work which might be considered perfect."

As early as 1720 operations with regard to the Nene outfall were commenced by Charles Kinderley, but owing to local opposition Kinderley's Cut was not finally complete till more than half a century later, *i.e.* 1775. Though an experiment on too limited a scale, it successfully established the principle that the chief obstruction to the discharge of the waters lay at the outfall.

In 1814 Mr. John Rennie was called in to advise. His recommendation involved a new channel "from the mouth of Kinderley's Cut to the level of low water in the bay," and other works for deepening the course of the Nene. Opposition due to financial reasons and to local feeling delayed any action till March 1826, when the North Level Commissioners undertook to carry out the necessary works "on receiving reasonable contributions from all the other parties to these measures."

Further works were authorised in 1830. The general effect of the opening of the Nene Outfall was most satisfactory. The scour of the river through the town of Wisbech added 10 to 12 feet to its depth and saved the town the expense of some £50,000, the amount estimated by Rennie in 1814 as required for deepening the river by manual labour.

Further important works in 1849 were carried out by Rennie and Stephenson.

The task which Fowler was called on to undertake in December, 1856, was the up-keep and improvement of this system of drainage. In a report addressed to the Commissioners of the Nene Valley Drainage and Navigation, Mr. Fowler pointed out that some of their works were in a dangerous condition owing to the excessive scour of the water in some parts of the channel and to excessive silting up in others.

A Bill authorising the raising of fresh funds had been rejected, and the commission was placed in a position of some difficulty. The engineer states, however, that as a temporary expedient he had placed across the river in the town of Wisbech, a contraction by means of a submerged weir and self-acting gates. This he had hoped to remove, when permanent alteration had been made, but he felt now obliged to fall back on the expedients authorised by their existing Act. The report, which is of great length, deals principally with matters of detail and contains a criticism of various proposals advanced in the interests of the different parties concerned in the drainage and navigation. These works, notably the temporary construction at Wisbech, caused great inconvenience to some, and the engineer emphasises the statement that the difficulty to be faced is not so much one of engineering as of reconciling a number of conflicting interests. The problem, in fact, was one to be solved by a development of jurisprudence as an antecedent to the operations of the engineer.

The Institution of Civil Engineers devoted more than one evening to the discussion of a paper on arterial drainage and outfalls, read by Mr. R. B. Grantham on November 29th, 1859.

The following is a résumé of Mr. Fowler's speech on that occasion. It may be mentioned that his intervention in the debates of the institution is very infrequent. He dwells, as is characteristic of his habit of mind, on the importance of detail and material, and on the danger, in engineering problems, of making hasty generalisations. Fowler's objection to generalisation is not that of the ignorant empiric, but of the man of experience, who knows the difficulty of apprehending all the factors of the problem about which generalisation has to be attempted. Without generalisation, science is impossible, but in applied science the factors of the problem are often so numerous and so complicated that generalisation may easily cease to be correct generalisation, and may fail, because it attempts to include, in identical propositions, facts and forces that are not homogeneous and which therefore require different formulæ for their explanation.

Fowler began by expressing his doubt as to the advantage of a commission with compulsory powers —the obstructiveness of private owners was a great evil, but possibly a commission might be a greater evil. He feared that a commission would not be competent from a professional point of view, and it might inaugurate works of a disastrous character on a large scale owing to the presence of some one member of great authority but no technical knowledge.

The remarks must be taken as a defence of the Nene Commission, which had no large compulsory powers.

With regard to arterial drainage he would, he said, "attempt to give direction to the discussion. The Ancholme drainage was a case where the outfall was not dependent upon the discharge of the water from the level itself. The Witham was a case where it was so dependent, but where the tide was shut out. The Ouse and the Nene were cases in which the outfall was dependent upon the discharge of the water from the level, but the tide was admitted. The Ancholme drainage . . . was one in which the tide was not admitted at all. The tidal current of the estuary of the Humber swept past its sluice entrance and kept it free, and it was not dependent upon the discharge of water from the land. The Ancholme was a combination of drainage and navigation with catch-water drains. The drainage was perfect as far as freedom from inundation was concerned, while the means of irrigation was afforded by the catch-water drains: and it combined better than at most other places drainage and navigation.

"Other cases of the kind had been treated upon the plan of Vermuyden, who, having a great dislike for tidal water, excluded it wherever it was practicable; and in some cases he was right. To the Witham, Vermuyden's principle was applied of shutting out the tide above Boston; but in that case it was wrong, because the outfall was dependent upon the mode in which the water was discharged, and consequently the navigation had been imperfect ever since. At Boston there was a large expanse of low land, over which the water struggled to get into the deep water of the Wash; the outfall, therefore, was dependent upon the way in which it was discharged. If the water was permitted to go up by the Witham in the same way as by the Ouse and the Nene, the Witham would be in a very different condition to that in which it was at present. The Ouse and the Nene might be termed open rivers, the Nene perfectly so, and the Ouse up to Denver sluice.

"The Nene was a river which had been really benefited by the alterations already effected. The outfall was very good, and all that was now wanted was to carry out improvements up the river; and by removing obstructions and properly attending to the banks, it would become a fine river. The Ouse was of a similar character to the Nene, and had fairly repaid all the care bestowed on it. Every improvement has been of great value both to navigation and drainage. Sir John Rennie had described the large amount of drainage that had been carried out in that district: Mr. Fowler, however, did not think that 13 feet had been gained at Lynn. (Sir J. Rennie later in the debate re-affirmed his statement, while Mr. Bidder, v.p., questioned Rennie's statement and supported Fowler.) Still, at the present time the Ouse below and above Lynn was a complete river. The cut obtained by Sir J. Rennie and the late Mr. Stephenson was certainly one of the finest engineering works in the country. The cut was two miles in length and from 600 to 700 feet in width, and a more successful result had never been obtained.

"He thought there was no branch of engineering science which required more care than such cases as these. The material was easily acted on by the scour, and even if cut originally in straight lines, it was easily distorted and turned into irregular forms. That required to be corrected, and it was the province of the engineer to decide upon the materials that should be used to protect the slope of the cut, upon the form to be given to the slope, and upon the means of maintaining the improvements at the smallest expense. The first depth obtained was seldom so much as was subsequently arrived at. These improvements were obtained at the risk of damage to the banks; as the depths increased, the banks must be protected, and constant watchfulness was necessary to carry them out with safety. In this respect it was impossible to lay down any general principle. The engineer must consider each case separately as it occurred to him, but as a rule a formation of rubble stone at the foot of the slope was a good system for directing the general current, and the upper part of the slope was kept in order with less difficulty. A minor improvement had been lately introduced into the works of the Norfolk Estuary by creosoting the stakes, and it was believed that this small additional expense was attended with advantage. He expressed his entire concurrence with the remark of Sir John Rennie, that it was impossible to lay down any general principle as applicable to all conditions and circumstances. The conditions of every case must be carefully and separately considered, and he specially cautioned engineers against the adoption of Vermuyden, of an inflexible principle in all cases."

Fowler's connection with the Nene Improvement Scheme was, even more than his railway work, a continuation of the labour of earlier engineers. As we have already pointed out, perhaps the most important

effects of the work of Fowler and his compeers in
connection with railway communications are the indirect economic results which, though each in itself
small, yet cumulatively amount to a great revolution.
So here, too, in addition to the question of arterial
drainage and navigation, and the reclamation of land,
the drainage of the Fens brings to light important
principles of economic progress.

The connection between the solution of engineering
problems and the development of the jurisprudence
of a civilised and progressive society is aptly illustrated
in the experience of the association known as the
Proprietors of the Norfolk Estuary, of which Sir John
Rennie and Mr. Fowler were the engineers. The tract
of swamp dealt with lay between the Nene Outfall and
the town of Lynn, and extended along the Wash, past
what are now the Sandringham estates of the Prince
of Wales. The inception of such work requires first
a clear definition of proprietary rights in respect of
the area to be treated, and then some reconciliation
of the conflicting interests of the various proprietors.
The following quotation from the engineers' report to
the proprietors for the year 1858 shows the method
followed in this instance:—

"During the past season the works of the Babingley enclosure,
recommended in our last report, have been completed, and about
550 acres of valuable land have been recovered from the sea.

"The value of the land was settled by arbitration before
being enclosed in the manner provided for by clauses in your
Act, and the increased value by reason of the enclosure is
now being settled in a like manner. When this has been
ascertained, the difference between the two amounts will represent the improved value of the land, and the Norfolk Estuary

Company will then receive one-third of it from the present proprietors, whose further rights as frontagers will thereby be extinguished, and in future all enclosures carried further seaward will be the sole property of the Norfolk Estuary Company."

Partly by aid of an Act of Parliament and partly by voluntary agreement, the proprietary rights in this derelict waste were defined. Without this security of tenure nothing could be done. This granted, contract and industry became possible, and in the present case the landlord on the existing foreshore paid to the adventurers one-third of the improved value of his reclaimed land and ceded to the adventurers all claim to further reclamations seawards. The engineering operations consisted of making a number of embankments generally in the form of wattled groins which afforded protection to land which was in danger from the scour of the tide, and which at the same time promoted an area of still water where the silt was so deposited as in time to render the area fit for enclosure. The process is known as " warping," and it is interesting to note that the visible sign that the land so reclaimed is becoming fit for enclosure, is the spread of the growth of the samphire plant. In the maps and plans of the reports the range of the samphire plant is carefully indicated.

We are tempted to add a further illustration establishing the converse proposition, namely, that the uncertainty of tenure which ensues in default of individual appropriation retards the beneficent action of the engineer and the agriculturalist.

The reward given to the original adventurers who drained the Bedford Level was certain farms which they redeemed from the swamp. Ownership, security of separate tenure, improved agriculture, and the maintenance of a sound system of drainage went together.

That part of the Fen which remained intercommonable—in other words, that which was every man's or no man's land—remained still a stagnant waste.

The particular piece of land which we have in view was an area "intercommonable of seven parishes" situate in the union of Ely, and bearing the unromantic title of Grunty Fen. A part of it was drained by the Earl of Bedford and his associates of the Bedford Level Corporation and reclaimed for the service of man. As time went on, however, the unreclaimed portion of Grunty Fen ceased to possess even the amenities, such as they are, of a good swamp. The drainage all round it rendered even the life of the pike precarious, and gave no safe breeding-ground for the wild duck. There was swamp enough left to give foot-rot to the sheep and establish ague in the shepherd's hut. The Fen was covered with ant-hills and thistles. Stray gunners poached at large after snipe. No one seemed to know who had legal rights on the Fen; everyone did what was right in his own eyes—dug it for sods, and even carried away the soil to the adjoining lands. A neighbouring owner, and, as such, possessed of commonable rights, describes how he set about abolishing this nuisance. "I ferreted about," he says, "in the Records of the Court of Exchequer and in the Petty Bag Office and ascertained what was the history of the other Fens before they were enclosed." As a result the proper procedure was ascertained, and a scheme for a *pro rata* allotment secured the requisite number of assents and an Act was obtained in 1861 for enclosure. The

interests of those who had rights in respect of the
Fen were adjusted by a valuer, lots were laid out,
roads and watercourses were made, and the recovered
acres were given over to separate ownership and cultivation. Something like £11,000 was expended in
the process mainly for drains and dykes and independent watercourses and outfall works.

"The crowning evidence of modern civilisation," says
Mr. Pell in his most interesting pamphlet (*The Making
of the Land of England*. Murray, 1899), from which
the foregoing particulars are taken, "is seen in a
railway bisecting the Fen with two stations on it,
bringing London within a two-and-a-quarter hours'
run." The story, which should be read at length in
Mr. Pell's pamphlet, is indeed an epitome of the history
of civilisation. It assigns with much precision the part
that is played both by the juridical reformer and by
the engineer.

The granting of a title to land in the hitherto derelict Grunty Fen enabled the adventurers to drain a
portion of it. The advantages of secure ownership and
the disadvantage of community of tenure being, in
process of time, made apparent, steps were taken to
end the ownerless condition of the remainder of the
fen. The land at length was given over to separate
ownership. This security encouraged industry, and
industry availed itself of engineering science, and
civilisation advanced. All this, gathered from the
microcosm of Grunty Fen, from the Norfolk Estuary
Association, and from the labour of its engineer, John
Fowler, must be weighed and considered if we would
frame aright the panegyric of the engineering profession.

CHAPTER VI.

METROPOLITAN RAILWAY

WE now come to what was probably the most extensive enterprise which Fowler carried through single-handed—the making of the Metropolitan Railway. The attention of the public is periodically directed to the growth of London traffic and the need of improved communications; and about the year 1853 the subject was being a good deal discussed. It occurred, of course, to many that some portion of the traffic might be carried underground, but it required great intrepidity on the part of a responsible engineer to undertake the making of such a line. Fowler, however, was nothing if not self-confident, and in 1853 we find him acting as engineer for the promoters of an experiment in this direction, and some description of this must now be attempted.

In 1834 there were steam carriages worked by steam power between Paddington (which twenty years before was described as a village four miles from London) and Moorgate, following very much the route of the present underground railway, but running on the surface.

After another interval of twenty years an Act of Parliament was procured for the " North Metropolitan Railway; Paddington to the Post Office; Extensions to Paddington and the Great Western Railway, the

General Post Office, the London and North Western Railway, and the Great Northern Railway. John Fowler, Engineer; John Hargrave Stevens, Architect." This Act of Parliament may be described as the first definite step towards the construction of underground Metropolitan railways, though the first portion of the then projected railway was not opened till thirty years later.

Apart from the question of facilitating the conveyance of passengers from one part of London to another, as early as 1845 the great railways were beginning to realise how important it was for them to have their stations as close as possible to their sources of traffic. Accordingly no less than nineteen bills were deposited in that session for lines within the Metropolitan area. A Royal Commission was appointed to inquire into the matter. Many projects were then put forward. Mr. Vignoles had a scheme for a Charing Cross to Cannon Street Railway. Messrs. Stephenson and Bidder were for extending the South Eastern to Waterloo Bridge. Mr. Locke suggested an extension of the London and South Western to London Bridge, and Mr. Page a line from the Great Western through Kensington to Westminster and along a proposed Thames Embankment to East Cheap and Blackwall. There was also a North London Railway and a Regent's Canal Railway. Mr. Charles Pearson, the City solicitor, was in favour of having a Great Central Terminus at Farringdon Street with a connecting line to King's Cross.

"Mr. Pearson," says the report of the commissioners, "would carry the line of railway between two rows of houses which he proposes to build so as to form a spacious and handsome street, 80 feet in width and 8,506 feet in length; the railway to be on the basement level and to be arched over so as to support the pavement of the street, which would be on the level of the ground-floor of the houses. Mr. Pearson proposed to give light and air to the railway by openings in the carriage-way and footpath."

Mr. Pearson, says Sir Benjamin Baker, from whose paper on "Metropolitan Railways" the information here given has been mainly compiled, "is clearly entitled to the credit of being the originator of the whole tribe of Arcade Railways." He was even the first to suggest "blow-holes," a subject which has occasioned much heartburning in subsequent years to the authorities charged with the maintenance of the public streets. In bearing testimony to the value of Mr. Pearson's services to the cause of Metropolitan communications, Mr. Fowler * remarked that his view was vitiated by the assumption that the public convenience required the concentration of railways in one central station, adding that if a man of the ability of Mr. Pearson had lived to see the great development of railway travelling, he would have abandoned his advocacy of concentration.

The Commission of 1846 was not, as Sir Benjamin Baker remarks, very far-seeing. They objected to the carrying of railway bridges across the river, as likely to create an obstruction to navigation; and none of the existing bridges could be given over to railway traffic without inconvenience to the public. They calculated that the average distance travelled by passengers arriving at the Euston terminus was 64 miles, and argued that the saving of another mile or

* *Minutes of Proceedings of the Institute of Civil Engineers*, vol. 81, p. 68.

two was of very little importance. The time was clearly not yet ripe for a Metropolitan system of railways.

Mr. Pearson, however, was not daunted, and continued, with his colleague Mr. Stevens, architect and surveyor to the western division of the City, to advocate various projects—for a City terminus line, for arcades, for vegetable and meat markets, and for Holborn Valley Viaducts. Finally, in 1853, an Act was obtained for a line of $2\frac{1}{4}$ miles, from Edgware Road to Battle Bridge, King's Cross. To this line Mr. Fowler was the engineer, Mr. Stevens the architect, and Mr. Burchell the solicitor. Plans were at once prepared for extensions both westwards to Paddington and eastwards to the City. The line authorised in 1853 was entirely under the public roadway, and it was not necessary to acquire any buildings, but the extensions were more ambitious, and it was proposed to tunnel under buildings and to acquire certain private property. In 1854 the Great Western Railway joined the promoters, and offered to contribute £175,000 of capital if the line from Paddington to the Post Office could be sanctioned. There was a severe parliamentary contest. Mr. Fowler was supported by Messrs. Brunel, Hawkshaw, Scott Russell, Peacock, and Stevens, and was opposed by many equally distinguished engineers. A great variety of objections were put forward. The tunnels could not be ventilated, the rails would be so greasy that no locomotive would be able to drag a train. This last objection was put forward by no less distinguished an authority than Mr. Locke, a great engineer, whom his countrymen have justly honoured with a resting-place in Westminster Abbey.

The approval of the Post Office authorities was at the time deemed of great importance. Sir Rowland Hill wrote to the chairman of the railway company :—

"The Postmaster-General thinks it necessary that such arrangements should be made as will admit of the intended railway being brought into the basement of the building, so that bags, when made up, may be placed at once in the railway carriages "; and this was part of the original scheme.

In 1855 a Select Committee of the House of Commons on Metropolitan communications was appointed, and Mr. Fowler appeared to give evidence as to "exact position of the Metropolitan Railway, for which an Act of Parliament has been obtained." The actual course followed by the Metropolitan Railway system has deviated so largely from the original design that it will be interesting to set down Mr. Fowler's evidence on this point :—

"In the session of 1853 a part of that scheme was sanctioned from near the Edgware Road to King's Cross, a distance of two miles, with a capital of £300,000. That was avowedly an imperfect scheme, and the intention was expressed of coming forward in the next session and completing it so far as to join the northern railways with each other and with the Post Office. That was done last session. The scheme of last session was to join the Paddington station by two lines—one for passengers and one for goods—in connection with the Great Western Company, who are shareholders in it, and to join the London and North Western at Euston Square, and the Great Northern at King's Cross, taking the line down to the Post Office, and there terminating in such a manner as shall communicate conveniently with the Post Office vaults, where the letters could be put into the vans. That was done in conjunction with the Post Office authorities, and with their entire sanction. The total length of that scheme, including the part sanctioned in 1853, is 4¾ miles, and the capital is £1,000,000. The works have not yet been commenced, owing to the war and the depressed state of the money market, and the determination of the directors not to proceed till the whole of the shares had been taken up. Every share has now been taken up, and the works are about to commence."

Such was the scheme as sanctioned by Parliament at that date. Mr. Fowler goes on to explain a further extension, "which will complete the communications with the Southern and Eastern Railways, both with each other and with the Post Office." The line was to communicate with the Eastern Counties Railway, with the Brighton, and the South Eastern, and to cross the river by means of a bridge. The addition in question was to amount to 3¾ miles. From the Post Office it was to run underground to near Old Fish Street, then to pass over Thames Street, and so over the river, and, by means of "a viaduct of ordinary construction," to join the South Western on one side and the London and Brighton on the other.

The extension to the Eastern Counties Railway was to be entirely underground, with a station at Moorgate Street in its course. This was an extension which the Eastern Counties authorities were anxious to secure, as they were dissatisfied with their existing station accommodation.

Opposite the Great Western Railway Hotel there was to be a passenger station, communicating by an under-ground passage with the "new" passenger station on the Great Western Railway.

There was also to be a "branch line," parallel to the Grand Junction Canal, for goods alone, and this was to be so laid out that a through train could pass along it quite conveniently. It was further in contemplation to make a similar goods connection with the London and North Western and with the Great Northern, " so that coal trains and goods trains can come along the Great Northern, and, by means of the North London, along the London and North Western to any part of the City, which the Metropolitan line will communicate with." This arrangement, it was pointed out to the witness, as at the moment authorised, unduly favoured the Great Western Railway. "It must be evident to you, Mr. Fowler," said the chairman, "that if this railway, which is to communicate with the north and south sides of the Metropolis, is ever sanctioned, it must be sanctioned upon the principle of giving to every railway company in the Metropolis the privilege of using it." Mr. Fowler's reply is characteristic of his method of meeting an objection. He adopts most enthusiastically the suggestion propounded by the adversary, and goes on to show how entirely it is met by the scheme which he is advocating.

It would have been a lengthy process to explain in answer to this question, as he does elsewhere, that "it is no part of their scheme to bring all the railways to one common centre. They do not believe that such a thing would be convenient, and therefore they have along the course of the line intermediate stations at convenient places"—convenient, that is, to the Metropolitan stations of the great lines. Mr. Fowler was all along of opinion that the Metropolitan Railway would be a passenger railway, and that its traffic would be of the "omnibus" character. This question about affording equal facilities for goods traffic to all the great railway companies was therefore of comparatively little importance, but it obviously took important rank in the mind of the chairman. The answer given is quite truthful and diplomatic. It agrees that equal facilities must be given, and "we are quite prepared to carry it out on that principle, so much so that I have considered the details of doing it, and it can be done quite conveniently." The question, however, really was whether the various companies would think the accommodation offered worth the expense which it would involve.

"The communications," Mr. Fowler added, "connecting the different railways can only be carried out in conjunction with those railways. The parties interested in the London Bridge Station and the South Western, as well as the great northern companies, must themselves feel the value of this communication, and must be active parties in co-operating to carry it out, in order to enable it to be done with prudence and success."

The conclusion of the committee is worth quoting :—

"On the preliminary point of their inquiry, your committee find that the requirements of the existing traffic of the Metropolis far exceed the facilities provided for it; that the rapid increase of that traffic is constantly adding to the amount of inconvenience and loss thus caused; that enormous as the increase has been, it is, and must continue to be, kept seriously in check by the want of means for its natural expansion; and that it has become indispensable to make provision in this respect for the future on a great and comprehensive scale, and with the least possible delay. In order that the House may the better appreciate the grounds on which your committee rest these convictions, the following facts, adduced in evidence by Mr. Charles Pearson, may be cited.

"The population of the Metropolitan District, which in 1811 was 1,138,000, by the census of 1851 was 2,362,000, thus showing that it has doubled in forty years. Your committee find that about 200,000 persons enter the City each day on foot by different avenues, and about 15,000 by the river steamers; and that besides the cab, cart, carriage, and waggon traffic of the streets, the omnibuses alone perform an aggregate of 7,400 journeys through the City. The number of passengers arriving at and departing from the London Bridge group of railway termini which in 1850 amounted to 5,558,000, had, in 1854, risen to 10,845,000. At the South Western Railway it appears that during the same period the numbers have increased from 1,228,000 to 3,308,000. The number arriving at and departing from the Shoreditch Station last year is stated at 2,143,000; Euston Square Station, 970,000; Paddington Station, 1,400,000; King's Cross Station, 711,000; the Blackwall Station in Fenchurch Street, 8,144,000. These figures serve to convey some idea of the rate at which the traffic of the Metropolis is increasing, and that increase, the House will bear in mind, takes place, notwithstanding the obstructions presented by the existing means of communication. Looking at the overcrowded state of the principal thoroughfares of a capital which, as has been shown, has doubled its population within the last forty years, and which there is every reason to believe is annually accelerating the ratio of its growth, your committee have arrived at the following conclusions. . . ."

The committee then sets out eight recommendations. The most important are to this effect :—

(3) That the different Metropolitan railway termini should be connected by railway with each other, with the docks, the river, the Post Office, so as to accelerate the mails, and take all through traffic, not only of passengers, but, in a still more important degree, of goods off the streets.

(4) That all existing restrictions upon the natural and convenient flow of traffic, such, for example, as tolls on the

roads and bridges within the Metropolitan district, should, as a general rule, be removed.

(5) That wherever, as, for example, in the case of the Metropolitan Railway, works for improving the communication of the Metropolis can be carried out by private enterprise, that course should be adopted, and your committee consider that most important results may be secured in this way.

(6) That when the improvements contemplated partake more of the character of a public benefit than of a commercial speculation, care be taken by such means as economising the waste spaces of the river, or opening new streets through poor neighbourhoods, to diminish the cost of the undertaking to the lowest possible point.

(7) Contains a recommendation for combining railway extensions with street improvements.

(8) That all the cost of public improvements required by the existing or prospective demands of Metropolitan traffic, whether in forming new streets, or enlarging existing ones, or in purchasing or building bridges, or removing toll-bars, should be defrayed by a local rate levied on the whole Metropolitan area.

The question of Metropolitan communications is, perhaps, as pressing to-day as it was forty years ago. The railways constructed by Sir John Fowler and his fellow-engineers have done much to relieve the congestion, but the economical difficulties which they partially overcame still remain, some of them, indeed, in an aggravated form. Towards private enterprise in the form of Metropolitan railways the committee adopts an attitude of benevolent neutrality. For foot-passengers and horse traffic the public authority should tax all the ratepayers, but there is no suggestion that the public authority should provide any sort of facility for station and line accommodation for the great modern convenience of railway travelling. It does not suggest, however, that this beneficent form of effort should be thwarted. This neutral, if not benevolent, attitude of the commission of 1855 has by no means been imitated. The whole system of imposing local rates on railways, their exclusion from any right of representation on local spending authorities, the passenger duty, the attempt to impose onerous conditions, in the way of rebuilding and workmen's trains, on the occasion of every application to the public authority, and, indeed, the whole attitude of Parliament to those public improvements which are carried out by private enterprise and "commercial speculation" seem to suggest that railways are a public scourge rather than a public convenience. Unfortunately for progress, the number of persons who are willing to deprive themselves and their neighbours of a great convenience, because someone may derive a commercial profit from its use, is largely on the increase. The result, of course, is to throw more and more into the hands of municipalism, that most important body of industrial enterprise which is obliged to invoke the aid of parliamentary powers for its inception. The suburban and metropolitan railway traffic of London is a monument of what private enterprise can do in spite of the opposition of popular misconception.

The ability of municipalism to conduct large enterprises may some day be proved. In the meantime its aggressive attitude has only succeeded in paralysing and obstructing private enterprise in such industries as housebuilding for the poorer class, the making of tramways, the extension of railway accommodation, the generation of cheap electrical power, telephones, and similar conveniences. Its ability to sustain by its own initiative the advance of economic improvement and to ensure the adoption of scientific invention for the benefit of mankind, which hitherto has been effected by private enterprise, is by no means apparent. Its period of usurpation is, however, too recent to enable men to see clearly what use it will make of the power which it has assumed.

This feeling of jealousy (ill-warranted as we believe it to be) against private enterprise working under compulsory powers is new. It has been the outcome of the comparative success of the railway and other similarly authorised undertakings. The difficulty to be overcome in the early fifties was of a somewhat different character, though it is all part of the same problem. The initial difficulty from which all others proceed is the necessity which *ex hypothesi* has arisen for the compulsory expropriation of private property for purposes of some public convenience. We have already given our reasons for considering that the right conferred on a railway company is by no means a monopoly in the sense that it enables the railway to charge exorbitant rates. This seems more especially to be true with regard to a Metropolitan railway, which is by no means set free from the competition of omnibuses, cabs, tramways, steamers, and (perhaps most important of all) of walking on our own ten toes. The whole endeavour of the owners of a Metropolitan railway must be to induce us to prefer their line to other methods of transport, and also to make journeys which, but for the convenience they offer, we should never undertake at all. The right to acquire property at an exorbitant price confers little benefit on the railway company. Justly enough, property acquired under such conditions has to be paid for at an ample rate of compensation, but this setting aside of the principle of free exchange, inevitable though perhaps it may have been, is the occasion of the whole difficulty of the situation. An owner who is compelled to sell naturally fights to obtain the utmost compensation; hence costly legal proceedings, and costly terms of acquisition. At the date of Mr. Fowler's first connection with the Metropolitan Railway no one ventured to say that the public would be overcharged, and that therefore the way-leave granted ought only to be granted to a public body. The main opposition came rather from property owners, and from those who alleged that in some way or other their proprietary interests were adversely affected by the proposed line. These, it need hardly be said, in a crowded Metropolis were innumerable, and when once the principle of free exchange has to be abandoned the difficulties of assessing values for the purposes of compulsory exchange become enormous. Indeed, considering the mass of property and interests affected, the business aspects of the construction of this line were really more formidable than its engineering difficulty.

In stating, in 1855, that the work of the Metropolitan Railway was about to begin, Mr. Fowler was somewhat too sanguine. The preliminary difficulties were not overcome till 1859, and the works were not actually commenced till March, 1860.

The original programme was varied from time to time. Sir B. Baker gives the following account of these changes of plan :—

"In 1861 powers were obtained for extending the Metropolitan Railway to Moorgate Street, and for widening the line from King's Cross eastwards; and in 1864 for constructing the Eastern and Western Extensions to Tower Hill and Brompton respectively, the District Railway from Brompton to Tower Hill and the St. John's Wood Railway. Owing to financial difficulties the Metropolitan Railway Company sought to abandon the Eastern Extension in 1870, but the Bill was thrown out by the Lords. An alternative mode of completing the Inner Circle by Fenchurch Street, instead of by Tower Hill, was authorised in 1871, but the latter and original route is the one adopted. As regards the Western Extension, Mr. Fowler's first idea was to take it through Kensington Gardens and Hyde Park; but the authorities objected, and the present line was selected. Practically there was very little choice in laying out the Metropolitan Railway, for the Lords' committee of 1863 decided that 'it would be desirable to complete an inner circuit of railway that should abut upon, if it did not actually join, nearly all the principal railway termini in the Metropolis, commencing with the extension in an easterly and southerly direction of the Metropolitan Railway, from Finsbury Circus at the one end, and in a westerly and southerly direction from Paddington at the other, and connecting the extremities of those lines by a line on the north side of the Thames.' The Inner Circle of railways as constructed is the direct outcome of that recommendation. The total length of the line is 13 miles 8 chains, of which about two miles are laid with four lines of rails, and there are twenty-seven stations."

The following is a table of the proportions made by the different engineers :—

ENGINEER.	LENGTH EXECUTED. Miles. Chains.		PERCENTAGE.
John Fowler	11	20	86
Edward Wilson		27	$2\frac{1}{2}$
Francis Brady		28	$2\frac{1}{2}$
Jas. Tomlinson, jun.		35	$3\frac{1}{2}$
Hawkshaw and Barry		58	$5\frac{1}{2}$
	13	8	100

Of the cost of this railway it is difficult to give any precise estimate, "because the financial liabilities assumed by the contractors varied from time to time, and other complicating conditions entered into the question." The following, however, is Sir B. Baker's summary on this head :—

"In 1871, when the works had been completed and opened from Moorgate Street to Mansion House, the expenditure on capital account by the District Railway Company for works and equipment of $7\frac{1}{4}$ miles of double line railway was stated in the Directors' Report to have been £5,147,000, and by the Metropolitan Railway Company £5,856,000 for $10\frac{1}{4}$ miles, both amounts being subject to deduction in respect of surplus lands. It must, however, be remembered that these figures, as already intimated, are of little value to an engineer, because they include items dependent, among other things, upon the market value of the shares, which in the case of the Metropolitan ranged from 50 to 140, and in that of the District from 20 to 100, during the making of the lines."

The completion of the so-called Inner Circle, according to Sir J. Wolfe Barry, was completed for about £3,250,000. This, however, includes something like £1,000,000 contributed to the making of new streets, as well as the cost of the Whitechapel extension.

It may here be noted that the so-called circle railway is not circular, but is really two parallel lines running east and west, connected at each end by comparatively short lengths running north and south. This fact, it is alleged, has interfered with the value of the railway as a communication between the north and the south. The set of the traffic of London has apparently always been from east to west. We may conjecture that this tendency was originally imparted to it by the waterway of the Thames, which, on the whole, runs from west to east. Every additional facility in the way of streets and railways increases the tendency of the traffic to concentrate itself on this route, so that it is difficult to say whether the growth of East and West London is due to comparatively ample transport accommodation, or whether the better transport accommodation is due to the agglomeration of population in the western and in the eastern parts of the city.

We have dealt with the route followed by the railway; we must next say something on the equally important question of the level on which the rails are laid.

"When a tract of country is closely covered by buildings of varying heights, and the natural watercourses are converted into covered sewers, it is difficult to form a general idea of the physical features determining more or less the character of the railway as regards level and gradients.

"In the case of the Metropolitan, however, a sufficient record exists of the previous conditions of the country, and the excavations have, in many instances, afforded an interesting confirmation of traditions."

The reader must be referred to Mr. Loftus, or some other history of London, for a picture of a primitive "Lynn din," or Port of the Waters, standing on the hill, which we now know as the City, with a tidal estuary running up from the Thames, along the course of the Fleet, or Hole-Bourne, a river which descended from the Hampstead and Highgate Hills. To the engineer the facts of the case are presented by the statement that within a few yards of their institution in Great George Street, Westminster, the level of the footway is eight feet below the highest tide, while at Hampstead the height is 443 feet above the ordnance datum.

"The highest ground traversed by the Inner Circle is at Edgware Road, and the lowest at the back of Victoria Street, Westminster, the respective heights being 103 feet and 8 feet above ordnance datum. At Swiss Cottage the rails of the St. John's Wood branch climb to a height of 167 feet, and at King's Scholar Pond Sewer in Victoria Street the District Railway dips to a depth of 9 feet below the same datum, or 3 feet below Thames low water."

The course of the streams that in olden times flowed to the Thames from the Northern Hills is always a matter of interest to Londoners, though now these streams are confined in sewers. Still, even in their degraded condition we are glad to hear of them from engineers who encounter them on their subterranean work.

"To the west of Kensington and Chelsea, rising in the high ground, but traversing chiefly the low-lying district, was the Bridge Creek, now known as the Counters Creek Sewer, which is carried under the District Railway at Warwick Road in a flat-topped channel 7 feet wide and 8 feet high. . . . Proceeding eastwards, the next stream met with was the West Bourne, rising on the western flank of Hampstead Hill and flowing southwards to the Serpentine, and thence into the Thames near Chelsea Bridge. This, now called the Ranelagh Sewer, is carried under the Metropolitan Railway at Gloucester Terrace and over the District Railway at Sloane Square Station, the construction in the former case being a brick channel 9 feet wide by 8 feet high, with flat iron top, and in the latter a cast-iron tube 9 feet in diameter, supported on wrought-iron girders of 70 feet span.

"Next in order was the Ty-Bourne, which flowed from

M

Hampstead through Regent's Park, thence by Marylebone Lane—whose strange windings are due to the houses having originally been built on the banks of the stream—and by the Green Park to the river between Vauxhall and Chelsea Bridges. At Baker Street the Metropolitan Railway is crossed by this stream under its present name of the King's Scholar Pond Sewer, the construction being a cast-iron oval tube resting on cast-iron girders. At Victoria Street the District Railway is similarly crossed, but the size of the tube is 14 feet by 11 in the latter case, as compared with half those lineal dimensions at the northern crossings."

The next stream to be dealt with was the famous Fleet, which proved the most serious difficulty in the construction of the railway.

"When building the retaining wall on the west side of Farringdon Street Station, the Fleet, then carried in a slightly-built brick sewer, 10 feet diameter, resting on the rubbish filled into the old channel, burst into the works and flooded the tunnel with sewage for a great distance. Again, when constructing the District Railway at Blackfriars the Fleet had to be diverted and re-diverted, carried temporarily in syphons, and otherwise carefully guarded, as the large volume of water coming down it necessitated. No less than five crossings of the Fleet had to be dealt with, namely, two at King's Cross over the junction curves with the Great Northern Railway, one at Frederick Street over the Metropolitan Railway, another at the same spot over the widening lines, and finally a fifth under the District Railway at Blackfriars Bridge. The first four crossings were in cast-iron tubes of tunnel section, ranging in size from 9 feet by 8 to 10 by 10 feet, and the latter in two brick channels 11 feet 6 inches by 6 feet 6 inches high, with flat iron tops.

"In regard, therefore, to the natural configuration of the ground it may be said that the railway on the southern side of its course ran along an old river-bed through what was once the swamps of Pimlico and Belgravia, while the northern section of the Metropolitan is on the lower slopes of the hills that lie to the north of London. The average rail-level of the District Railway is 13 feet below Thames high water, while the northern section is 60 feet above the same level.

"For the purpose of joining the high levels of the north with the low levels of the southern lengths of the railway, deep cuttings or tunnels were necessary in the eastern and western parts of the line.

"In construction, cuttings 42 feet deep and a tunnel 421 yards in length are found at Campden Hill on the west, and cuttings 33 feet deep and a tunnel 728 yards in length at Clerkenwell on the east; the respective gradients to get down the sloping ground being about ¾ mile of 1 in 70 on the west, and 1 mile of 1 in 100 on the east. Further, as the valleys of the West Bourne and the Ty-Bourne are crossed by the Metropolitan Railway, dipping gradients of 1 in 75 and 1 in 100 are required at these points. On the District Railway the rise from the Fleet Valley to the hill upon which the earliest parts of the city were built has to be surmounted, and it is done by a gradient of about ½ mile of 1 in 100."

Contrary, therefore, to what we might have expected, the levels and gradients of the Metropolitan Railway have been very little affected by the buildings and underground communications of London; they have followed very closely the natural configuration of the land exactly as they might have done in an open country. The curves, on the other hand, which have a minimum radius of 10 chains on the Inner Circle and 6⅔ chains on the Great Northern branch, were fixed chiefly by the situation of the sources of traffic and by property considerations.

In his address to the British Association in 1882 Mr. Fowler made the following comments on the difficulties of working a railway with frequent stoppages, and throws out a hint as to what would be an ideal arrangement of gradients. The Great Northern Railway, he points out, runs its trains from London to Grantham, a distance of 105 miles, without stoppage, so that little power and time are lost. On the Metropolitan, on the other hand, no sooner has a train acquired a reasonable speed, than the brakes have to be sharply applied to pull it up again.

"As a result of experiment and calculation," he says, "I have found that 60 per cent. of the whole power exerted by the engine is absorbed by the brakes. In other words, with a consumption of 30 lbs. of coal per train mile, no less than 18 lbs. are expended in grinding away the brake blocks, and only the remaining 12 lbs. in doing the useful work of overcoming frictional and atmospheric resistances.

"Comparatively high speed and economy of working might be attained on a railway with stations at half-mile intervals, if it were possible to arrange the gradients so that each station should be on the summit of a hill. An ideal railway would have gradients of about 1 in 20, falling each way from the stations, with a piece of horizontal connecting them. With such gradients, gravity alone would give an accelerating velocity to the departing train at the rate of 1 mile per hour for every second; that is to say, in half a minute the train would have acquired a velocity of 30 miles an hour, whilst the speed of the approaching train would be correspondingly retarded without the grinding away of brake blocks. Could such an undulating railway be carried out, the consumption of fuel would probably not exceed one-half of that on a dead-level railway, whilst the mean speed would be one-half greater. Although the required conditions are seldom attainable in practice, the broad principles should be kept in view by every engineer when laying out a railway with numerous stopping-places."

Most interesting are some of the historical and geological details which the construction of the railway brought to light. In some places the engineers found themselves working through 24 feet of ruins and dust, the deposit of bygone generations of Celts and Romans, Saxons, Danes, and Normans. The foundations of an old fort were exposed at the mouth of the Fleet, and at the Mansion House a masonry subway was discovered intact. Along the old river-bed the foundations had to be sunk in some places to a depth of 37 feet below high water. Beds of peat were encountered in the swamp of Pimlico. Below the peat is the London clay, and on this the railway walls between Gloucester Road and Victoria were made to rest. On the top of the clay were found varying depths of sand and gravel heavily charged with water. This fact necessitated an elaborate system of pumping, costing, it is estimated, about £600 per month, during the period of construction. Permanent pumping-stations are now established at South Kensington, Victoria, Sloane Square, and the Temple stations.

The north section of the line was driven through gravel and sand beds of comparatively small dimensions, lying on the top of the London clay.

The cuttings and longitudinal section of the railway show that if what Sir Charles Lyall called the great ochreous gravel deposits of the Pleistocene Age were swept away, the hills and valleys of the Metropolitan area would be practically unaltered in appearance. At what period in the remote past the sand, gravel, and brick-earth, cut through by the railway in Westminster at a level of 8 feet and in Marylebone at 103 feet above ordnance datum, were deposited no one can tell. Geological speculation, however, suggests that England was then united to the Continent, that the present site of the North Sea was dry land, and the Thames a tributary of the Rhine, along whose banks disported themselves mammoths, woolly rhinoceros, and other extinct animals, and, possibly, also man.

The Underground Railway consists of covered ways, tunnels, and open cuttings with retaining walls.

The greater portion of the line is what is technically called a covered way, that is to say, the buildings and the roadway were removed and the necessary excavations made; the railway was then laid, and again covered in, and the buildings in some cases reconstructed. A variety of expedients were employed. The type of covered way adopted in the first portion of the railway consisted of " a six-ring elliptical arch, of 28 feet 6 inches span and 11 feet rise, with side walls three bricks thick and 5 feet 6 inches high from the rails to the springing." Originally there was no " invert " for the support of the side walls, but experience seems to have shown that this was desirable, and in later constructions this was generally added.

In places there was not sufficient depth for a brick-covered way, and here iron girders were substituted. The cast-iron girders appear to have been more satisfactory than wrought-iron, as the former suffered less from oxidation; but Sir B. Baker's verdict is that the brickwork, though more costly and involving deeper cutting, has proved, in the long run, the cheaper and more satisfactory form of construction.

The erection of the covered way was attended with the greatest difficulty. Temporary roadways had to be made in order to interfere as little as possible with the traffic of the streets. Considerable improvements and

THE MAKING OF THE METROPOLITAN RAILWAY.

Marylebone Road (looking East).

To face page 106.

economies were effected as experience dictated. They
began apparently by excavating the whole width of
the cutting, and then protecting the sides and the
foundations of the contiguous buildings, during re-
construction, by balks of timber and struts of a more
or less elaborate character.

The following is Sir B. Baker's description of the
latest and most ingenious device :—

"In later days, when the District Railway was being constructed, the general practice was not to timber the entire
width of covered way, but to sink a couple of six feet wide
trenches for the side walls, to build the latter up to four feet
above springing, take out the excavation full width down to
that level, fix the centering, turn the arch, and finally take out
the 'dumpling.'"

By this contrivance the risk of subsidence of the
adjoining land was reduced, and the new permanent
supports were introduced before the whole of the old
supports were finally removed.

Although to the ordinary traveller the whole of the
Underground seems to be in tunnel, technically there
are only three tunnels on the line—the Clerkenwell
tunnel, 728 yards long, on the original Metropolitan ;
the "Widening" tunnel, 733 yards in length, parallel
to the preceding ; and the Campden Hill tunnel, 421
yards in length. Considerable amount of trouble was
experienced with regard to these tunnels, more especially at Campden Hill. The two first-named tunnels
were driven for the most part through hard and dry
clay, and, as a matter of mere tunnelling, were comparatively simple.

"However," says Sir B. Baker, "with the utmost precautions
tunnelling through a town is a risky operation, and settlements

may occur years after the completion of the works. Water mains may be broken in the streets and in the houses, stone staircases fall down, and other unpleasant symptoms of small earthquakes alarm the unsuspecting occupants."

In the driving of the tunnels, as well as in making excavations for the covered way, it was necessary to proceed by very short lengths at a time in order to lessen the risk of landslips. Some difference of opinion arose between the engineers and the contractors on this point in connection with that most troublesome bit of the line which traverses Campden Hill. The soil here is extremely light and loose, and the engineers insisted that it should be worked in six-feet rather than twelve-feet lengths.

With regard to that portion of the line which is made in open cutting, no special remark has to be made, except that in open cuttings excavated in the neighbourhood of buildings more than ordinary precautions are necessary. The same plan which has been described with regard to the covered ways was followed here. At first the whole width was timbered and excavated. In the later work trenches were sunk and the retaining walls built, and then the "dumpling" or centre portion was removed in railway waggons.

Such was the normal course of construction. There were, however, innumerable difficulties of a special character to be overcome. It must be remembered that forty years ago work of this class was in its pioneer stage.

"It is now known," says Sir B. Baker, "what precautions are necessary to ensure the safety of valuable buildings near to the excavations; how to timber the cuttings securely and keep them clear of water without drawing the sand from under the foundations of adjoining houses; how to underpin walls, and, if necessary, carry the railway under houses and within a few inches of the kitchen floors without pulling down anything; how to drive tunnels, divert sewers over or under the railway, keep up the numerous gas and water mains, and maintain the road traffic when the railway is being carried underneath; and, finally, how to construct the covered way, so that buildings of any height and weight may be erected over the railway without risk of subsequent injury from settlement or vibration."

It was originally intended to work the line by means of locomotives of special construction. Very light trains of three carriages, running at intervals of five to ten minutes, stopping at alternate stations, would, it was thought, be sufficient. Mr. Fowler proposed to have a locomotive using no fuel, but having simply a reservoir of hot water capable of being heated up again at the end of each journey, and a reservoir of cold water to condense the steam; still he maintained that the line could be worked by ordinary locomotives. Brunel supported this view strongly, and declared that even with ordinary locomotives no special means of ventilation would be found necessary. There had been a standing order in former days that required committees of Parliament to report on the means of ventilation proposed for tunnels. Engineers now thought that this was unnecessary, and in proof Brunel instanced his own Box tunnel, where a screen had to be introduced because the draught was excessive. The trains were to be worked by hot-water locomotives, were not with passengers to exceed 20 tons, and the trip from Paddington to the City was to occupy from twelve to fifteen minutes. Sir B. Baker, who takes a special delight in pointing out the frequent futility of the prophecy even of experts, remarks that the hot-water engine was not even tried, trains of 120 instead of 20 tons are used, and the time is doubled. The earlier part of the line was constructed with a view of accommodating traffic of small dimensions, and it is not to be wondered that the ventilation was not altogether satisfactory. The experiment of the hot-water type of engine had to be abandoned when it was decided to allow the trains and engines of other lines to have running powers over the Metropolitan. The earlier portion of the line was, for the convenience of the Great Western Railway, laid with broad-gauge rails, and the earliest engines used on the Metropolitan were supplied by the Great Western.

The type of engine is thus described in Sir B. Baker's often-quoted paper :—

"It was a six-wheel broad-gauge tank engine, having four coupled wheels 6 feet in diameter, and outside cylinders 16 inches by 24 inches. The heating surface of tubes was 615 square feet, of fire-box 125 square feet, and the grate area was 18·5 square feet; a tank of 375 gallons capacity and a condenser of 420 gallons were provided."

This arrangement, after continuing for some months, was ended by a dispute as to terms between the Great Western Railway and the Metropolitan. The Great Western Railway withdrew their engines and the Metropolitan had to turn for assistance to the Great Northern Railway.

"The engines used were six-wheeled tender engines, having four coupled wheels 5 feet 4 inches in diameter, cylinders 15½ inches by 22 inches, and a total heating surface from 760 to 940 square feet. Experience showed that both the Great Western and the Great Northern engines were too light for their work, and the powerful tank engines now used on the line were designed by Mr. Fowler and Messrs. Beyer and Peacock. These engines have four coupled wheels 5 feet 9 inches in diameter, a four-wheel bogie, and cylinders 17 inches in diameter by 24 inches length of stroke. The heating surface of the fire-box is 103 square feet, of the tubes 909 square feet, and the grate area is 19 square feet. The weight in working order is between 42 and 43 tons, with 1,000 gallons of water in the tanks."

Mr. Fowler, who was prevented by illness from being present at the reading of the paper from which we have so largely quoted, wrote a letter which is entered on the minutes, stating how grateful he was " to his old friend and partner, Mr. Baker, and he was sure every member of the institution would be equally so, for the trouble he had taken in searching through old documents bearing upon the early history of the Metropolitan Railway, and describing so fully the works of the Inner Circle." He associated himself with Mr. Baker in his tribute to Mr. Pearson, the City solicitor. The only other point he desired to mention was one which might be encouraging to the younger members of the profession. At the monthly meetings of the directors, members of the Board used frequently to tell him that they were warned by engineers that the line would never be made ; that even if made, it could never be worked ; and that even if worked, that no one would travel by it. This, as he mildly puts it, was rather discouraging, but he knew his business and believed in it, and the directors believed in him, and he did make it, did work it, and the public travelled by it.

The first section of the Metropolitan Railway was

opened on January 9th, 1863. The idea that the
Metropolitan was a line of connection between the
great trunk lines having stations in London is to be
found in most of the notices of the opening of the
line. The *Times* describes the new line as the beginning of a system of conveying traffic across London
without break of carriage, an expectation which we
know has never been realised.

An entry made by Mr. Fowler in his wife's diary
records that Professor and Mrs. Owen came to Queen
Square Place, by appointment, at twelve o'clock, and
accompanied the engineer, Mrs. Fowler, and their eldest
son to Bishop's Road Station, where they received the
Lord Mayor, Lord Harris, Mr. Lowe, and other distinguished guests. At one o'clock they started in the
first train. They stopped at each station and examined
it thoroughly, especially those at Portland Road and
King's Cross, till they reached Farringdon Street.
Part of the station here was enclosed, and an "elegant
déjeûner" provided by the directors for 650 people.
The Prime Minister, Lord Palmerston, had been invited
to take part in the opening ceremony, but excused himself on the score of his age and engagements, adding
with a sparkle of his never-failing humour, that for his
part he was anxious to keep above ground as long as
he could. In his absence, Mr. R. Lowe, M.P., called
on the company to drink a bumper to the success of
the Metropolitan Railway, and described Mr. Fowler
as the modern St. George, who had four times van-
quished the Fleet ditch. The enterprise, he declared, "was an honour to the country, and a solid
advance worthy of civilisation." The Lord Mayor con-
gratulated himself and the Corporation because they

A TRIAL TRIP.

METROPOLITAN RAILWAY, EDGWARE ROAD STATION.

1. Lady Constance Grosvenor.
2. The Duke of Sutherland.
3. Mrs. Gladstone.
4. Mr. T. M. Johnson, *Resident Engineer*.
5. Mr. George Knight, *Contractor*.
6. Mr. Charles Gilpin, M.P.
7. Right Hon. Stuart Wortley.
8. Lord Wenlock.
9. Sir Stephen Glynn, Bart.
10. Mr. Charles Wood.
11. Lord Richard Grosvenor, M.P.
12. Lord Macclesfield.
13. Lord Grosvenor.
14. Lord Ronald Gower.
15. Mr. Fowler, *Engineer*.
16. Right Hon. W. E. Gladstone, M.P.
17. Mr. Armstrong, *Contractor's Agent*.

To face page 172.

had relaxed their rule against embarking in private speculation, and had allowed themselves to subscribe £200,000 towards the great public improvement which they were inaugurating. Lord Harris proposed the health of Mr. Fowler, who in reply gave a brief history of the construction of the line. The difficulty of raising capital he ascribed mainly to the Russian War, and he concluded by publicly thanking those who had been engaged with him in carrying out this great work.

"Of the directors it would not be proper for me to say more than that they have always given me a most generous and unbounded confidence, without which I feel the task would have been beyond my powers. I wish especially to thank my old assistant and friend, Mr. Johnson, who has acted as resident engineer of the line throughout, and of whose assiduity and intelligence I cannot speak too highly. I have also to thank the contractors, Messrs. Smith and Knight and Mr. Jay, for the admirable manner in which they have done their work and their prompt attention to my instructions."

Among the distinguished men who showed interest in the new line none were more eager than Mr. Gladstone, and, shortly after its opening, Mr. Fowler sent him, in April, 1863, an elaborate statement of its progress, from which the following extract is taken :—

"From the great interest you took in this work during its construction, it has occurred to me that you will feel an interest in its working; and in the hope that it will not be troublesome to you, I have had a few statistical facts prepared for your perusal. The line was opened to the public on January 10th last.

"The number of passengers already carried
exceeds 2,000,000
The average receipt per passenger is . . 2·8224*d*.
The average number of passengers daily is 28,687
Greatest number carried in one day . . 60,000

"The following figures give the mileage, passenger receipts per mile, and total number of passengers in one year of all the Metropolitan railways :—

Name of Line.	Mileage.	Total Number of Passengers in 1862.	Total Number of Passengers per Mile per Annum.	Passenger Receipts per Mile for One Week in March.
				£ s. d.
Great Eastern . . .	664	10,375,322	15,625	16 14 2
Great Western . . .	992	7,302,156	7,361	22 2 9
London and North Western	1,179	18,142,506	15,388	29 11 3
London, Chatham and Dover	72¼	2,508,175	34,715	29 13 6
Great Northern . . .	330	5,237,323	15,871	30 1 9
London and South Western	442¼	8,925,987	20,183	30 16 7
London, Brighton and South Coast . . . . }	247¼	11,928,355	48,244	40 2 1
South Eastern . . .	306	12,388,227	40,484	45 8 11
North London . . .	9	6,580,173	731,130	216 15 7
Blackwall	5¾	9,103,390	1,583,200	247 0 0
Metropolitan according to present rate of traffic . }	3¾	10,470,750	2,792,200	629 11 5

"Although the traffic of the Metropolitan line has been very large, it must be remembered that the City terminus of the company is at present situate in Farringdon Street, a distance of about one mile from the Bank. When the company's extension to Finsbury is completed, the number of passengers will be largely increased.

"The traffic of the line at present is purely local, and is entirely of the character of an omnibus traffic. There is a large traffic between the terminal stations, but a very considerable proportion is conveyed between intermediate stations.

"The proportion of passengers of each class carried in the month of February was as under:—

First class . . . . 17 per cent.
Second class . . . . 31 „
Third class . . . . 52 „

"The experience of working for some months has quite dispelled all fear as to noise and vibration to either streets or houses. I have heard no complaints of any kind, and the feeling generally appears to be that a vast convenience is accomplished without interferences with streets or otherwise, and that the appearance of London is not prejudicially affected by anything we have done.

"Altogether it is a source of great satisfaction to me that the result appears fully equal to the opinion you were kind enough to form of the interesting undertaking. I know how much you are engaged, and especially at the present time, and I beg you will not take the trouble to acknowledge the communication.

"I remain, my dear sir,

"Yours very truly,

"JOHN FOWLER.

"The Right Honourable W. E. Gladstone, M.P.

"*April*, 1863."

Among the numerous letters of congratulation received by Mr. Fowler on the completion of his great enterprise, none was more valued by him than that which was sent to him by his father (then in his seventy-ninth year), who, notwithstanding his age, had come up from Sheffield to attend the ceremony :—

"WADSLEY HALL,

"*January* 12*th*, 1863.

"MY DEAR JOHN,—I had a comfortable journey home on Saturday, and arrived safely at Sheffield at two o'clock. Although I had no companion on my way, my thoughts on the opening of the Metropolitan Railway fully occupied me.

"I most sincerely congratulate you on that occasion. Your numerous friends felt a gratification on the proceedings of that day no less than your father, although none could feel as your own father could feel, having had the opportunity of watching from your youth that progress in life which, from your natural talent, united with great energy and sound judgment, has been the means of raising you to the high estimation in which your friends and the public hold you.

"The proceedings of last Friday will be long remembered by your friends, and they will look back on that day with great pleasure, and none more so than your own father, while life is permitted him to enjoy the reflection of that day's proceedings. . . . I have had the opportunity of reading the report of the proceedings (which I think correctly stated), and especially your speech; all parties who have read it seem to make out that it was to the point, and the proper word, and each word in its place.

"Your generosity in speaking of Mr. Johnson's great assistance does credit alike to your head and your heart. Several who have read your speech make the same remark as myself.

"And now, my dear John, I hope that the very high estimation in which your friends and the public hold you, will not have the effect to raise yourself nor any part of your family in your own importance, but that you will keep that humble and even path in life which has hitherto marked your progress in all these matters. I beg to thank you for your kind and pressing invitation to the opening, and I am glad I accepted it. I believe I have not taken any cold, and am very well this morning. . . .   "Yours affectionately,

"JOHN FOWLER.

"John Fowler, Esq."

A correspondent, Mr. Charles Stanley, writing congratulations to Mr. Fowler on the opening of the Metropolitan Railway, recalls a committee-room incident of the year 1845 :—

"During the great parliamentary campaign of 1845, when two rival lines were projected from Sheffield into Derbyshire, I remember hearing you cross-examined as to one of them which skirted the moors. 'I believe, Mr. Fowler,' said the opposing counsel, 'your line goes by the name of the *Grouse and Trout Line.*' 'Very likely,' you replied, 'and your line is called the *Flute Line,* because it is nearly all tunnel.' It is singular that you should have now sensibly added to your fame by the construction of a line, the distinguishing characteristic of which is that it is nearly all tunnel—in fact, a genuine flute line."

With regard to the services of the late City solicitor, Mr. Pearson, Mr. Fowler was much gratified in being able to put in a favourable light the claims of his widow on the Metropolitan Company and also on the Corporation, and among the letters preserved is a very grateful acknowledgment of his kindness from Mrs. Pearson.

The work of Fowler in making the Metropolitan Railway was a pioneer effort, and many improvements have since been added to the business of tunnelling under cities. The usefulness of a pioneer effort is often, as we have already insisted, to be measured by the rapidity and completeness with which its methods are improved and superseded.

In the foregoing description of the Metropolitan Railway we mentioned how important the engineers considered it to complete the tunnel arch at as early a point of the construction as possible. In the later period of the work the arch was keyed and the soil under the arch, "the dumpling," was removed afterwards. In recent years the shield patented by Brunel as long ago as 1818 has been used with success for the driving of iron tunnels through the bowels of the earth at so low a depth that disturbance of the surface and buildings no longer occurs. A steel cylinder forced forward by hydraulic pressure prepares the way, and, as the material is removed and the steel cylinder works its way forward, an iron tunnel of a slightly smaller diameter is put in its place. The space so left round the outside of the iron tunnel is filled with a preparation of strong cement, forced in by compressed air at a pressure of 50 lbs. to the square inch. The whole structure is thus sealed and completed with a minimum of disturbance to existing buildings.

N

One other element, denied by a series of accidents to Fowler's earlier work, is necessary to the successful use of the shield system of deep tunnelling. Mr. Fowler, as we have seen, abandoned very reluctantly his special engines designed for the use of the Underground Railway. In his address to the Merchant Venturers' School, in 1893, after reciting the circumstances under which this took place, he adds :—

"But the ventilation has always been an objection and a difficulty, notwithstanding various attempts to improve it; but experience has shown that so long as trains run every two or three minutes, no satisfactory ventilation can be obtained with the present type of engine, or any modification of it, and the only solution of the difficulty with our present knowledge appears to be the adoption of an electric motor similar to that used on the South London Railway.

"In my opinion we may expect henceforth that all underground railways in London will be constructed on the plan of the Oxford Street scheme, which avoids disturbing the streets or houses above the railway, and worked by an electric motor."

The late Mr. J. H. Greathead was the engineer, and Sir John Fowler and his partner, Sir Benjamin Baker, were the consulting engineers on the City and South London Railway, which was the first railway constructed as above described.

CHAPTER VII.

PRESIDENT OF THE INSTITUTION OF CIVIL ENGINEERS

IN 1863, as already narrated, the first portion of the Metropolitan Railway was completed. The most interesting excursion of the following year, 1864, was the inspection of the Glasgow Waterworks. The bursting of a reservoir at Sheffield had caused a good deal of alarm among the persons responsible for waterworks throughout the country. Mr. Fowler had been employed in the inquiry which followed this disaster. With him were Mr. La Trobe Bateman and two other engineers. They unanimously agreed that the bursting of the reservoir was due to the sliding of the clay foundation of the dam and puddle trench (under the pressure of a landslip in an adjacent hill) upon the shaly rock beneath.

The Glasgow municipality very naturally desired to reassure themselves by submitting their works to the inspection of the best engineering opinion of the day. Accordingly, in August, 1864, Mr. Fowler, accompanied by the Lord Provost and the committee of the waterworks, Mr. Bateman and Mr. Gale, the engineers, and other gentlemen, made a tour of inspection among the beautiful lakes of the Trossachs which supply Glasgow with water. The works, he writes to his wife, are

"well and neatly executed," but, he adds in his
capacity of salmon fisher, "steps or stairs have been
made for the passage of salmon from the river into the
loch above (Loch Vennacher), but they are not
executed in a proper manner," a point which, we are
glad to find, was duly noticed in his formal report
to the Corporation. The works on Loch Vennacher
and Loch Drunkie are only incidental to the supply of
Glasgow, and were really obligations put on the
Corporation for the benefit of the mill owners on the
streams below.

The water-supply of Glasgow, as all the world
knows, is taken from Loch Katrine, by means of
sluices placed on the south side of the loch. These
sluices admit the water into a tunnel and aqueducts
which carry it to Glasgow, a distance of about thirty
miles.

"The works of the Glasgow Corporation Waterworks, which
we inspected yesterday," he writes to his wife, "consist chiefly
of a tunnel under the hills, 8 feet wide and 8 feet high, with
aqueducts over the valleys constructed of stone piers about
45 feet apart, with a wrought-iron trough to convey the water
resting on them. Some of these works are of considerable
extent. The larger valleys, of which there are three, are
crossed by pipes laid down to the bottom of the valley and
up the other side to save the great expense which would have
been involved by long, high aqueducts."

The report to the Corporation on the state of their
property, dated November 7th, 1864, is signed by
John Fowler, J. F. Bateman, and James M. Gale, and
is very satisfactory and reassuring. Certain minor
recommendations involving an expenditure of about
£9,000 were made.

"We have given our reasons in detail for the recommendation of expending a sum of £9,000, and have stated that a
considerable portion of this expenditure is rather to place
every portion of the work, as far as human foresight can do
so, beyond the possibility of serious casualty than to remedy
existing defects; but our report would scarcely be complete,
or convey a correct impression, unless we were to add, in
conclusion, that the works are admirably fulfilling the purposes
for which they were designed, as to present and future requirements; and the great works of construction are of the most
sound and durable character."

This approval of the Glasgow Waterworks was of
considerable importance, for Glasgow has been a
pioneer in the movement which has led great centres
of population to bring their water-supply from large
natural or artificial lakes, situated at a distance.

Here, again, the art of the engineer is raising new
and important questions of equity and jurisprudence.
To whom does the water-supply of an upland country-
side of right belong? What compensation is to be
given to landowners, to other towns competing for the
same source of supply, to country villages and farms
whose springs are tapped by the marauders from the
great towns? The Glasgow water-supply was, of
course, taken from an already existing lake. By comparatively simple means the level of the loch was
actually raised, notwithstanding the water taken from
it, and so comparatively little disturbance was made ;
but when artificial lakes are made, and the springs of
many hillsides are diverted into aqueducts for the
supply of a distant town, the case becomes more
serious ; and, if we mistake not, we are here face to
face with one of the most difficult questions of the
immediate future. The problem has been raised by

the new values created by engineering progress, and, as on so many other occasions, we have no doubt that the progress of jurisprudence will be found equal to its solution.

Later in the year we find he was in the Isle of Wight, superintending the extension of the island railways, the greater part of which had been laid down by him.

In 1865 he took the important step of purchasing from Mr. Davidson of Tulloch the Highland estate of Braemore, and from that date he began to identify his interests with Scotland.

Of a visit paid to Dunrobin about this time, for the purpose of advising the Duke of Sutherland about his private railway venture, he writes an interesting account to his wife. " It is," he says, " a most difficult decision for the Duke to make, but I hope to be of use to him." The result of the Duke's deliberations on the question of the railway was that he made a considerable portion of this line at his own sole expense. This was ultimately sold to the Highland Railway at a price very much below what it had cost the Duke to make. In the evening a novel employment was found for Mr. Fowler. One of the ladies of the party was prevailed on to sing, if he would finish her letters dismissing one cook and engaging another. This he accomplished most successfully, and, he says, in exactly the words which were thought to meet the occasion. He also was commissioned to write a letter to an absent husband. In this, however, he was less successful, as his effort was pronounced to be " not quite affectionate enough."

On February 13th, 1865, Mr. Fowler became a member of the Engineer and Railway Volunteer Staff

Corps which was then being organised. In this body he continued to take a warm interest for the rest of his life. He was made commandant of the corps in 1891. This interesting and useful body never, we are informed, makes any public appearance as part of Her Majesty's forces, but it is a body to be consulted by the War Office in matters in which it may be thought advantageous to obtain the opinion of railway experts. In December, 1892, he received the rank of honorary colonel and the Volunteer medal given in recognition of twenty years' service.

The years from 1860 to 1866 were probably the busiest years of Mr. Fowler's life. The office entries show that he and his staff were during each of these years actively engaged in advising upon, and in many cases actively superintending, on an average, some seventy or eighty large engineering undertakings. It is not too much to say that at this period no considerable engineering project was set on foot without his service being called in, either for the promoters or for the opposers. The following are among the principal works belonging to the decade 1860–1870 : extensions of the Metropolitan Railway to Farringdon Street, the Metropolitan District Railway and other extensions of the underground system, the Oswestry and Dolgelly line, the Bristol and Clifton Railway, Liverpool Central Station, the Glasgow and City Railway, with a viaduct over the Clyde, St. Enoch's Station in the same city, Millwall Docks, railways in Devon and in Cheshire. He also advised H.M.'s Office of Works as to the bridges in the Regent's Park and as to the Serpentine and other ornamental waters in the London Parks.

In 1866 he was elected president of the Institution

of Civil Engineers, the highest honour which it is in the power of his professional brethren to bestow. He was at this time only forty-nine years of age, and the youngest president who ever sat in the chair. His portrait was painted by Millais, and presented to the Institution by his fellow-engineers. A reproduction of this, one of Millais' earliest portraits, is added to this volume. His presidential address is so characteristic of the man and of the position he occupies in the history of engineering that we make no apology for reproducing it in full. It is mainly concerned with the education of the engineer.*

"We of the passing generation," said the president, "had to acquire our professional knowledge as best we could." It is characteristic of Fowler that, though his education was of the most practical nature, no one was more keenly alive to the advantage of applying sound scientific principles to engineering work. He supplied the deficiency of his earlier training by an astonishing quickness in apprehending the teaching of the new science, by the aid of precedents gathered in a professional experience of extraordinary variety, and by the careful selection of qualified assistants.

The *Times* of January 11th, 1866, makes the following comment on engineering education. The sentiment reads a little strangely at the present day. The views of Mr. Robert Lowe (Lord Sherbrooke) as to educational endowments were still a power in the columns of the *Times* and in the land. After dwelling on the dangers of endowments, "which, if they encourage and direct, also have a tendency to narrow and confine," the article continues :—

* See Appendix.

Sir John Fowler. Bart. K.C.M.G.
from the portrait by Sir John E. Millais P.R.A.
in the Institution of Civil Engineers.

"Civil engineering has hitherto been free from this danger. If endowed, it is at any rate free. Offering very lucrative prizes to the ambition of the students, it leaves them quite at liberty to seek for the instruction necessary to obtain these prizes when and where they please. The result has been what might naturally be expected; the human mind, being steadily bent on attaining one particular kind of knowledge, has found the means of knowledge for itself, and the engineers of England occupy a position all the prouder because it is due to no extraneous assistance.

"Such were the feelings with which we heard the excellent address to the Institute of Civil Engineers from their president, Mr. Fowler, a man who though not, as far as we are aware, possessed of any university degree, or any other distinction than the modest letters 'C.E.' which follow his name, has probably done as much to promote the prosperity and spread the reputation of England as any person of his class in the country."

The following retrospect from an address delivered by Sir John Fowler more than a quarter of a century after this date is worth quoting as showing the sustained interest which he took in engineering education :—

"It is now twenty-seven years ago," he says, addressing the Merchant Venturers' School at Bristol in 1893, "when, as president of the Institution of Civil Engineers, I selected for the subject of my inaugural address the education and training of the engineer. . . . The time had evidently arrived for taking up this subject with earnestness and energy, and wide circulation was promptly given to this address in England and our colonies, and especially by the Indian Government, who sent out copies to all their engineers in India.

"We had already gone far beyond the stage of 'rule of thumb,' when good work and good materials were considered sufficient to constitute excellence. It was inevitable, under such a rule, that much of the material got into the wrong place, or was wasted in needless profusion. In those days of 'rule of thumb,' when a well-known engineer was remonstrated with on the excessive quantity of material he put into his works, he made the reply, 'Remember, it takes rather a clever man to know when a work is too strong, but any fool can tell when it falls down.'

"The word 'practical' is a word of which English people have always been proud, and no doubt it is an excellent and significant word; but before the comparatively recent introduction of technical training, sound theories, and careful calculations the expression 'practical' applied only to a thorough knowledge of the quality of the material and workmanship, and not to their economical and scientific arrangement. "But now the word 'practical' would be felt to be almost a term of reproach unless it was combined with a knowledge of sound principles.

"On the other hand, the word 'theory,' or 'a theoretical person,' were terms almost of derision, and no doubt when you saw a young student with just sufficient mathematical knowledge to calculate the stresses on a simple girder, dubbing himself an engineer, without the slightest knowledge of the quality of the material or of the workmanship, you were not surprised to hear such a person spoken of with considerable contempt. . . .

"This one-sided knowledge is now a thing of the past, or very nearly so, and the professor's knowledge of theory and the workman's knowledge of practice are expected to be combined and taught together to the same individual, both in regard to large things and small. . . . The rate of progress in arriving at the present stage of theoretical and practical teaching has, of late years, been astonishingly rapid, and especially since about the date of my address in 1866. . . . It is common knowledge that the facilities were not then either great or numerous; and although it would be unbecoming and incorrect to say that there were no good technical colleges, yet I suppose for one technical institution in those days there are now fifty, and the higher qualifications of the teaching staff and the superior character of the laboratories have progressed almost as much as the numbers of the institutions."

His interest in engineering education led him to support and follow with friendly sympathy the fortunes of the Engineering College at Cooper's Hill.

Speaking at the distribution of prizes on July 25th, 1888, Sir John told his audience that his first connection with the college had been at the time when the Government of India found themselves unable to give employment to every Cooper's Hill student who had passed his examination, as had been the practice, and when it became a question whether the college should be continued or given up altogether, Lord Cranbrook, who was then Secretary of State, asked him for his advice.

"I then," continued Sir John, "took on myself the responsibility of advising that the experiment of keeping on the college should be made, provided that the Government of India felt themselves at liberty to give the students a certain limited number of appointments in India. This has been done, and the result was most thoroughly and completely successful."

During the last years of his life he was consulted by a friend as to the best education for a son who intended to follow engineering as his profession. Sir John's reply was at once shrewd and practical. "If the young man has the force of character to resist the temptations to idleness which beset the student at the Universities, and if he has any aptitude for a serious study of the higher mathematics, then let him go to Cambridge." Otherwise, Cooper's Hill and the earliest possible introduction to the practical work of the profession were his recommendations. The conditions on which a Cambridge curriculum was considered desirable were very stringently expressed, so much so that probably very few young men would be able to satisfy them. Such at least was the conclusion of both father and son in this particular case.

In the year 1870 Mr. Fowler visited Norway as member of a commission appointed to advise the Indian Government on the subject of gauges. A broad gauge of 5 feet 6 inches had been adopted in the great trunk lines of India, but for the less important lines it was thought that economy and efficiency would be served by the adoption of a narrower gauge. Mr. Fowler's experience in respect of the Great Western line had formed in him a strong objection to any break of gauge. The question submitted to the commission, however, assumed that the Government of India felt itself obliged to adopt a break of gauge, and the experts were asked to say which narrow gauge was the best. Mr. Fowler's colleagues were General Strachey, Colonel Dickens, and Mr. (now Sir) A. Meadows Rendel. The Duke of Sutherland accompanied the party in his private capacity, and the commission was attended by Mr. Carl Pihl, the Norwegian Government Engineer of Railways. The Norwegian railways are on a 3 feet 6 inches gauge. The rails and engines are very light, and the rate of speed is very slow. Mr. Fowler thought the gauge was narrow enough, and that the rails and engines were too slight for economy. His colleagues, however, recommended a 2 foot 9 inch gauge as suitable for India. Mr. Fowler wrote a minority report to the effect that the 3 foot 6 inch gauge was the narrowest which he could approve. In the event the Government has adopted the metre gauge, a 3 foot $3\frac{1}{3}$ inch gauge, thus inclining more towards Mr. Fowler's opinion than to that of the majority report. In 1889 he paid a visit to India, and his inspection of the Indian railway system seems to have satisfied him that his advice was correct.

The bent of his mind in this, and indeed in all other matters, was ever against what seemed to him a "penny wise and pound foolish" policy. During this visit to India in 1889 he was consulted by Sir F. (now Lord) Roberts, on the question of the steep gradients on the Jhelum and Rawalpindi Railway, which varied from 1 in 50 to 1 in 100. The alternatives proposed for improving this most important strategic line were : (1) to double the existing line without altering gradients—marked by Fowler in memo. now before us as "unanimously rejected"; (2) to use heavier engines and rails—rejected as impracticable owing to sharp curves and liability to breakdown due to enormous engines, and strain on couplings, etc. ; (3) to regrade the line, leaving two stiff gradients of 1 in 50 where bank engines should be employed. This last proposal had been provisionally accepted by the Government, and an expenditure of some 18 lacs of rupees sanctioned. The objections to this were the expense of bank engines and the inefficiency of the arrangement for press of traffic in war times. Mr. Fowler estimated the cost of regrading the line throughout at 49 lacs. His characteristic conclusion is as follows :—

"As a practical question for a commercial railway it would, in my opinion, be wise to spend 31 lacs to get rid of two banks of $\frac{1}{50}$. If so, still more would it be wise to do so to meet the possible military requirements, when a breakdown (probable enough with the hurry and confusion and rush of war emergencies) would be an incalculable risk. I cannot doubt the Indian Office will sanction the improvement of the whole line to $\frac{1}{100}$."

The experience which he gained, and the work which he did in these years, seem to justify the estimate of his qualifications as an engineer which he himself propounded, when in the year 1882 he was invited to preside over the Mechanical Science Section of the British Association, which met at Southampton.

" A well-informed man," he said, "has been defined to be a man who knows a little about everything and all about something.   If you give me credit for being a well-informed engineer, I will endeavour to justify your good opinion by showing, whilst presiding at these meetings, that I know a little about steam-navigation and machinery generally, a little about steel and iron and other manufactures, and I trust a good deal about the construction of railways, canals, docks, harbours, and other works of that class."

This, we are inclined to think, is a very just estimate. What Sir John did not know about the construction of railways, canals, docks, and harbours was probably not worth knowing.   He followed, moreover, the results of improved processes of manufacture; and knew exactly how to turn them to engineering uses.   He had the eye of a skilful general thoroughly understanding the objective of his campaign, able to use to advantage every arm of the service, and well aware of the necessity of securing for his lieutenants the best scientific intellects of the profession—in a word, a great organiser, with sufficient scientific knowledge to make his organisation thoroughly efficient.

As Fowler has himself told us, M. de Lesseps was at pains to explain to his friends that he was not an engineer, yet the Suez Canal will in all future ages be associated with the name of Ferdinand de Lesseps, and rightly so. Fowler, unlike Lesseps, was a competently trained engineer ; yet, if we are not mistaken, they both owe their great positions to the same qualities. Sir John Fowler's achievement and the important place which he undoubtedly occupies in the engineering history of the nineteenth century, are due not so much to his high scientific acquirements as to his practical powers of organisation. In framing the panegyric of Sir John Fowler, there is no need to depreciate the value of correct theory. No one has insisted more emphatically on the paramount importance of scientific training, but the impartial bystander will not fail to notice how frequently high scientific attainment is combined with a certain futility of purpose, and how requisite to the business of this industrial age is the organising mind. If we wished to claim a scientific pre-eminence for Sir John Fowler, it would be a pre-eminence in that most difficult and as yet unformulated science, the management of men and of finance. This quality of intellect is necessary for successful achievement, and, in its highest forms, it is certainly rare. Mere scientific knowledge of the brute forces of nature is by no means rare, it can be hired in many markets on honourable yet by no means extravagant terms.

Sir John Fowler is a type of a class of men whose importance is apt to be under-estimated. They move through life in comparative silence, they make themselves no great addition to our knowledge, but if we may use the hackneyed words in this connection, *si monumentum quæris circumspice.*

The subdivision of labour in the service of mankind has proceeded far. Invention is one thing, the utilisation of invention is another. To be inventive and to be practical is not always given to one and the same person. We have all heard and laughed at the story of Lord Ellenborough, who removed his banking account when he heard that his banker (Samuel Rogers) wrote poetry, but there is a certain justification of this view to be found in the sober reality of history. The purely inventive and scientific genius is not always the safest guide, though undoubtedly its work is essential to human progress. We need not go to the length of Lord Ellenborough, who declared his intention of seeking out the stupidest man in the world of banking as the fittest to be entrusted with the safe custody of his cash, but we do obviously require the sifting process of stolid, unimpressionable common sense, if the world at large is to enjoy the full benefit of the progressive victories of science. The selection of what is practicable, and the rejection of what is impracticable, are functions which require very high qualities, and every day they are becoming more imperatively necessary. This is the useful work which is performed by the class of men of whom Sir John Fowler was an eminent type.

CHAPTER VIII.

MR. FOWLER AND GENERAL GARIBALDI
AN EPISODE

AN interesting but not important episode in Mr. Fowler's life may now be narrated. It will serve to mark the fact that he now enjoyed a European reputation, and will display his diplomatic adroitness, a quality to which he owed much of his success.

The inundations of the Tiber in the Roman Campagna and in Rome itself had for long been a subject of anxiety for Italian statesmen. Disastrous floods had occurred in the year 1870, and numerous projects were put forward for the reclamation of the Campagna and the relief of Rome from the danger of floods and malaria.

Naturally, the standing, perennial problem of the Italian engineer interested Fowler. When visiting Rome in 1871 he discussed the question with Count C. Arrivabene. This gentleman, under date June 19th, 1873, sent him, by the hand of Baron French, the head of a well-known financial firm in Rome, a proposal for getting up a company, of which Mr. Fowler was to be the leader, for the drainage of the Campagna. Nothing appears to have come of the matter, but the subject became urgent when in 1875, after the entry of the Italian Government into Rome, General Garibaldi took his seat in the chamber as a deputy, and announced his intention of devoting himself to this great national work. The task which had baffled the power of imperial and papal rulers naturally challenged the energy of the democratic liberator.

The General's advent in Rome was not without its anxiety for the government of Victor Emmanuel. It was so far a matter of congratulation that the hero renounced all intention of making experiments on the constitution, but it was soon found that his schemes for social improvement, though benevolent and magnificent, were not always consistent with sound and business-like engineering. His plans, indeed, were at first on the largest and most ambitious scale. Like all true visionaries, the great Italian patriot prided himself on being a practical man.

"I am only a practical man," he told the correspondent of an English newspaper. "I am resolved to push forward the project, as soon as it is approved, into immediate action. What I want is that a beginning should be made, and if I only see a trench opened I shall know that the work is begun."

There is no more dangerous character in the world than the popular hero declaring with his hand on his heart that "something must be done," and with only a very vague notion as to what that something ought to be. The General's authority with the people of Italy was of course unbounded. His behaviour to the monarchical authorities had been magnanimous, and the government was bound to treat his proposals with the greatest deference.

The engineering problem may be stated briefly as follows. The Roman Campagna is a large triangle with its base, some 88 kilometres in length, resting on the sea. The apex of the triangle is stretched towards the Apennines, and consists of a surface of about 1,000,000 acres; a great part of this area is uninhabitable for many months of the year. The Tiber winds through this plain and overflows into the swamps and marshes. The whole region is only about nine feet above, and some of it even below, the sea level.

The ingenuity of Mr. Lecky has discovered that the malarial dangers of the Roman Campagna have conferred on Italy one great boon, namely, that its parliament can hold no prolonged session, owing to the unhealthiness of the summer season; but otherwise the swamps of the Campagna have been considered deplorable evils. This condition of things, moreover, was obviously capable of improvement by wise engineering measures.

The plan of the General is thus described by a well-informed correspondent. It was to divert the waters of the Tiber into the bed of the Anio for about a couple of miles, as far as the Nomentano Bridge, and then to cut an entirely new and navigable canal, which would serve also as a collector to the whole system of lateral drains down to the sea. The old bed of the Tiber was, in the General's first conception, to be filled in, though later he proposed that it should still be preserved, controlled by locks which should regulate the flow, and enable the antiquary to explore the bed of the river for the lost treasure of earlier times.

There is, perhaps, nothing impracticable in all this, except the question of cost. This to the "practical man" who wished something to be done was a detail, but it was a consideration not to be ignored by Signor

Minghetti, the Prime Minister, and his financial advisers.

The idea seems to have been that the engineering works to be agreed upon should be carried out by an association of adventurers, whose remuneration should consist in grants of reclaimed land, and in sums raised in the form of taxation or loans by the Government and by the municipality of Rome.

Among the persons interested were the Duke of Sutherland, the Prince Torlonia, and other distinguished capitalists. At this point Mr. Fowler was asked to intervene, first by the Duke of Sutherland, whose friendship with Garibaldi had induced him to interest himself in the problem, and then by the Italian Government. Sir John Fowler's account of the transaction may be given in his own words :—

"Before leaving Cairo I received a telegram from Rome to the effect that a serious difference of opinion had arisen between the Italian Government and Garibaldi with reference to the mode of dealing with the River Tiber, so as to prevent a repetition of the injury to health and property which had resulted from a recent flood; and I was invited by both parties to go to Rome and endeavour to reconcile the differences.

"On my arrival I called on M. Minghetti, then Prime Minister of Italy, to obtain information regarding the question upon which my assistance was requested. He informed me that the matter was very serious indeed, for Garibaldi's position at that time was one of great influence, and it would be dangerous to have a quarrel with him, and yet at the same time it was impossible to entertain his views.

"I then called upon Garibaldi, who at once said, 'I am a pessimist; the Tiber is a danger to Rome, and therefore I say remove the Tiber.' This was rather startling, and I explained that being an engineer, I could only give an opinion on facts, and that I should first desire to ascertain what would be the cost and consequence of removing the Tiber as he proposed from Rome to Tivoli. Garibaldi assented to this as being reasonable, and I consequently lost no time in obtaining engineers from the Italian Government to make surveys and sections over the ground, and an estimate of the cost.

"This was done as rapidly as possible, and I called upon Garibaldi to state the result of my investigation of his suggestion, which practically involved a cost of about nine millions sterling and an equal amount for compensation. I went through the details with him, and he frankly admitted that I had demonstrated the impracticability of his scheme, and he was much obliged for the trouble I had taken."

Sir B. Baker, who was present at some of the interviews of his chief with General Garibaldi, has described how they were received by the Italian patriot. His young wife, like a peasant girl, sat in the room sewing, but took no part in the conversation. The General, who received all and sundry, was much exposed to tourists and adventurers of all kinds. He said to Fowler, jokingly, that he was much taken with the proposal of some American engineers, who were prepared to deviate the Tiber for no other payment beyond the antiquities which they expected to discover, " while you," he said, turning to the English engineers, " say the thing will cost millions." Mr. Fowler succeeded to some extent in inducing the old man to hold aloof from the irresponsible crowd which he admitted all too readily to his confidence.

The General took a very gracious leave of his visitors, and presented Fowler with his photograph and autograph.

Before leaving Rome Mr. Fowler addressed the following letter to General Garibaldi by way of summing up the situation, and with the purpose of at least delaying the General's decision. It displays, we venture to think, considerable diplomatic skill on the part of the writer, and is a good instance of Fowler's adroitness in the management of men.

"ROME, *March* 28*th*, 1875.

"DEAR GENERAL GARIBALDI,— Before leaving Rome for England to-morrow morning, I wish to send you a very short note of the result of my visit to Rome.

"Our excellent friend, the Duke of Sutherland, communicated to me in Cairo by a telegraph from Rome that he thought I could be useful to you in your great undertaking, and that you would be very pleased to see me on the subject.

"It has been a great gratification to me to come to Rome and to see you and examine this great question.

"At the special wish of the Duke of Sutherland and the Duchess, I called upon Sir A. Paget and S. Minghetti.

"I found Minghetti's disposition excellent. He told me he had no wish or views as to the mode of accomplishing your objects, that he entirely sympathised with them, and was prepared to give hearty assistance to any proposal which did not exceed a total amount which did not embarrass the finances of Italy.

"He immediately gave me all documents in his possession which supplied information, and expressed a hope that the Duke of Sutherland would be induced to co-operate with you in your work.

"Having the advantage of your general views and the documents supplied to me, I next examined the valley of the Tiber, the marshes on each side of the Tiber, and the site of the proposed harbour, in which I was accompanied by Mr. Wilkinson.

"My previous knowledge of Rome and the neighbourhood, and my attention having been called to sanitary Roman improvements in 1871 and 1873, greatly facilitated my present proceeding.

"Since the consolidation of Italy into one kingdom, and the adoption of Rome for its capital, the rectification of the Tiber and the diminution of malaria have become not only an Italian but an European necessity.

"The result of my careful studies during the present visit is, that rectification of the Tiber and the improvement of the marshes on each side, which you have taken up with so much enthusiasm and general sympathy, are perfectly practicable and reasonable work, and may be adequately supported with prudence by the Italian Government.

"The rapid fall and physical conditions of the Tiber are favourable to works of rectification for the prevention of injurious floods in the city.

"The remedies for this evil naturally divide themselves into two parts.

"1st. A diversion of the Tiber from the city.

"2nd. An improvement of the Tiber in and near the city.

"The object desired can be accomplished by either of these means.

"Until the profile or section was prepared of the diversion of the Tiber from Rome, no opinion could be formed upon it.

"This section has been prepared during the last few days, and has been furnished to me. The work of excavation involved is great, but it presents no peculiar difficulties beyond its magnitude.

"No borings have been taken, and therefore an approximate estimate only can be made at present. I greatly fear, however, that under the most favourable circumstances the total cost of such a diversion, including the filling of the old channel, drainage, purchase of property, and expenses consequent upon its construction would be beyond that which could be prudently entertained.

"I do not think it possible that this total cost can be less than from five to six millions sterling.

"Of the reputed advantages and disadvantages of the removal of the Tiber from Rome, in respect to its navigation, current of air from the mountains, and otherwise, I will not speak, because on such subjects others are better, authorities than I am.

"Fortunately you are not dependent for a satisfactory solution of the question upon a complete diversion. The documents placed at my disposal, and my own examination, show me that by means of a small diversion of the river below Rome and proper works of rectification in and above Rome, your object can be adequately accomplished.

"Of the value of drainage and cultivating the marshes on each side of the Tiber I have the highest opinion. The work is simple and easy, and certain in its results of improved value and sanitary condition.

"Of the harbour—I see no objection to Mr. Wilkinson's plan of breakwaters; it would, however, be prudent, in my opinion, to have boreholes put down through the alluvial deposit of the Tiber on the site of the proposed breakwater, so as to estimate with greater accuracy the probable settlement of the blocks.

"The marshes should be drained and made healthy before the port is opened for business, or its character as a safe and proper position would be at once destroyed.

"I have no statistics to enable me to form an opinion of the commercial necessity for a harbour, but those will be in your possession.

"On the whole, my dear General, my suggestion would be to have the studies of the diversion completed, including borings and other information, and a trustworthy estimate prepared by competent Italian engineers on whom you and others can rely, discarding those who would mislead you by imaginary quantities and prices.

"If the result of this investigation produces much more favourable features than my approximate estimates present with existing information, then extend the inquiry into all its consequences and have full designs and estimates prepared.

"If, on the contrary, you find the cost and consequences to present insuperable obstacles to its realisation, let a complete scheme of rectification be prepared of the existing Tiber through Rome.

"In either case I would advise that the work of improving about 100,000 acres of the Tiber marshes be included in the project of Tiber rectification.

"I am of opinion that a satisfactory plan of Tiber rectification, on the basis of preserving the course of the river through Rome, including the reclamation of the marshes I have described, could be carried out for a total cost of two millions sterling, and within a period of three years.

"Allow me, in conclusion, my dear General, to express my gratification in seeing you upon this great and useful question, and to assure you that the Duke of Sutherland and myself will always be happy to render you any assistance in our power.

"Believe me always,

"Yours very truly,

"JOHN FOWLER."

Though leaving Rome in March, Fowler was kept advised as to negotiations. A trusted member of his staff, Mr. W. Wilson, was deputed by him to report upon the various schemes which were brought forward. In a letter to Mr. Fowler, written in May, 1875, Mr. Wilson describes his labour in making himself conversant with something like a dozen rival schemes.

Garibaldi had apparently got the estimate made as Fowler had recommended. The General, Mr. Wilson reports, had yielded to the feeling of the people of Rome, who objected to a diversion of the Tiber from the city, and, indeed, was altering his plans from day to day. A new plan received by Mr. Wilson the very morning of his report proposed a complete diversion from the junction of the River Anio, round the eastern side of Rome, joining the Tiber again near St. Paulo, but the old course was to be preserved, and the flow of the river was to be regulated by locks. The flood-water was to run down the course of the new diversion, while the river in its ancient channel would be maintained at its ordinary level. The cost of this was estimated by Garibaldi and his advisers at 75 million francs. In Mr. Wilson's judgment it would cost at the very least 100 million francs.

Both of these sums were beyond the amount which the Italian Government was prepared to advance, and the scheme itself was open to many serious objections. The principal of these was, that, at normal times, a division of the water of the river would not insure the proper scouring of both channels, and would in fact aggravate rather than improve the insanitary condition of the river.

Fowler's success in inducing the General to condescend to estimates, secured a delay and saved the State from the undue precipitancy of the hero.

The sequel seems to have been that a commission examined some nineteen plans, rejected those which involved a deviation of the river, and selected one which proposed to clear the existing channel from numerous obstacles, such as old bridges, piers, and other superfluous masonry, to wall and embank the river within the city to a height of 55 feet, to rebuild certain bridges with larger openings, to give to the river a minimum width of 109 yards, to construct main drains or sewers on each side of the embankment. This plan had the approval of Mr. Fowler. The Government adopted the recommendation, and the works were commenced in 1876.

CHAPTER IX.

A PROPOSAL FOR A CHANNEL FERRY

M R. FOWLER, as we have already seen, became a pioneer in the matter of Metropolitan railway communications, and, as we shall have occasion to relate, he was the chief engineer of the Forth Bridge, a structure which entirely revolutionised the railway communications of Scotland.

All Mr. Fowler's projects, however, did not end in success, and the following narrative of his failure to revolutionise the passage to France will not be without its interest.

The history of the International Communications Bill, by which it was sought to authorise a Channel Ferry, deserves for its own sake a somewhat full notice. Though rejected for the time being, it still remains the most feasible of the methods proposed for facilitating the journey from England to the Continent.

The Bill was brought before a committee of the House of Commons on April 29th, 1872, and naturally Mr. Fowler, as the engineer of the scheme, was a principal witness in its favour.

We condense the following account of the proposal from the evidence which he offered in its support. The subject of better communications with the Continent, he said, had engaged his attention from 1864 onwards, and in the following year, 1865, surveys of the coast of France and England were made by himself and by his able coadjutors, Mr. Abernethy and Mr. W. Wilson, with a view of selecting a suitable place for the accommodation required. Dover was selected as the best starting-point on the English side, but the promoters could not at this time come to terms with the Admiralty, who objected to certain interferences with the Admiralty pier. The negotiations were renewed in 1867, but without success. In the meantime Mr. Fowler was applied to by persons who proposed a tunnel.

"I came to the conclusion," he said, "that at all events it was premature. Yes, and I declined to take the responsibility of adopting it. I thought it very much better that we should begin with something which at all events we could see our way to the end of. It may lead, in the course of fifty or a hundred years, to a tunnel being seriously proposed, but we are certainly not ready for a tunnel yet." "The bridge," he goes on to say, "is too ridiculous to discuss, as the bridge would consist of a number of piers, which would be rocks dangerous to navigation."

There remained, therefore, nothing but his own plan, which he shortly described :—

"My proposal is parallel to a tunnel in this sense, that it is a continuous communication. The very essence of my proposal is that carriages, goods trucks, and mails should be carried across without breaking bulk; that would be accomplished by a tunnel, but it will be accomplished much better, in my opinion, by proper boats."

The proper boats which he had in view were boats of about 450 feet long, of 6,000 tons burden, and 10,000 horse-power, capable of carrying a train of sixteen carriages, the weight of which was estimated at a little

MODEL OF THE CHANNEL FERRY STEAMER

To face page 205

over 100 tons. The carriages were to be run bodily
on to the ferry boat and conveyed "without breaking
bulk" across to France. The railway companies
already possessed the necessary powers for building
such boats, but parliamentary powers were required
for alterations in the harbour at Dover and for the
construction of what he called a water station. The
details of the changes at Dover Harbour need not be
particularly described, as operations conducted by the
Government on a much larger scale are now in progress.
The general result was that a new harbour of 95 acres
was to be obtained, which would secure perfectly still
water for the use of the ferry boats.

The water station was to be fitted with hydraulic
lifts, wharves, and the necessary apparatus for the
passage of the trains to the steamer. The hydraulic
lift was to be one half the length of the vessel, and
trains of the London, Chatham, and Dover, and
South Eastern Companies were to be put on the lift
as they arrived, and run on board the boat side by
side. Sir William Armstrong had been consulted
about the machinery, and undertook to provide lifts
that would complete the operation in a minute. Mr.
Fowler, more cautiously, thought that five minutes
would not be an unreasonable delay. Some extension
or alteration of the companies' lines was necessary for
the project. The London, Chatham, and Dover were
at first favourably disposed, but the South Eastern,
whose interests were identified with Folkestone, and
with the tunnel scheme of which its chairman was
the principal promoter, were strong opponents of Mr.
Fowler's scheme. The Harbour authorities of Dover
also opposed before the committee of the House of

Commons, but their opposition was withdrawn when the Bill reached the Lords.

Satisfactory assurances of support had been received from the French Government through M. Thiers, and there seemed reasonable ground for hoping that the concession necessary for constructing the corresponding landing-stages on the French coast would be given. Mr. Ward Hunt, a leading member of the Conservative party, and some time First Lord of the Admiralty and Chancellor of the Exchequer, gave evidence that in the year 1869 he and Mr. Fowler had an interview with the Emperor of the French, and received considerable encouragement. War broke out between France and Germany in 1870, and negotiations came to an end. Favourable assurances had also been received from M. Alphonse de Rothschild, chairman of the Great Northern Railway of France. The negotiations were renewed, through the instrumentality of Mr. Philip Stanhope, one of Mr. Fowler's assistants, in 1871, when M. Thiers, the President, more or less directly renewed the favourable assurances of the French Government. He stipulated, however, that the necessary concessions for beginning the work on the English side of the Channel should first be obtained before it would be possible for him to approach the French Chambers on the subject. A variety of considerations seemed to point to Andrecelles as the place of debarkation for the ferry steamers on the French coast. From the nature of the case, therefore, the promoters were unable to produce any definite agreement binding the French authorities to accept and forward their proposals.

The estimated cost of all these works was stated by Mr. Fowler to be £890,000.

To make a long story short, the committee of the House of Commons found the preamble of the Bill proved, but added the condition that the powers applied for should cease and determine if within a given time (two or three years were mentioned as the proper limit) the necessary concessions were not obtained from the French Government.

In July of the same year the Bill was brought before a select Committee of the House of Lords, consisting of Earl Belmore (Chairman), Earl of Moreton, Earl of Ilchester, and Lords Raglan and Lawrence. Much the same ground was travelled over. It appeared, however, that in the interval the South Eastern Railway had to some extent succeeded in detaching the London, Chatham, and Dover from their support of the Bill. It was also brought out that, in the early stages of the promotion, negotiations had taken place, and, as it was hoped by the promoters, an agreement reached between the International Communications Company and the authorities of the South Eastern Railway. The arrangements, however, broke down on the question of terms, and the promoters' counsel made much of the point that practically both the companies had agreed to the principle of the Bill, and that their opposition, as now shown, was merely prompted by the question of terms. Against the Bill it was urged that the promoters

"stand entirely alone, without friends in the English railways, without friends in the French railways, and without any assurance from the French Government that the conditions of their undertaking will be satisfied, also without any prospect of any person whatever finding them the capital necessary for this undertaking."

In vain the counsel for the promoters drew glowing pictures of a brilliantly lighted water station, invalids in search of warm climate conveyed to the sunny south without risk of exposure from the weather, the abolition of sea-sickness, the vast increase of traffic, and improved harbour at Dover for craft of all kinds, and the *entente cordiale* between our island and the Continent strengthened and confirmed. In the ruthless official language of the report, the committee-room was cleared; after some time the counsel and parties were again called in, and the chairman, Earl Belmore, announced that

" the committee have given a careful consideration to this Bill, which they consider one of considerable importance, and the majority are of opinion that the preamble is not proved."

The value of all this " consideration " is not perhaps enhanced, when it is added that the committee was evenly divided, and that the adverse decision was adopted on the casting vote of the chairman.

Two coincidences which arose in the course of the evidence, personal to the subject of this biography, are worth noting. One was the evidence given as to the steam ferry which conveyed goods trains across the Forth from Granton to Burntisland, a service destined to be more or less entirely superseded by the Forth Bridge. Secondly, the evidence of Mr. Robert George Underdown, general manager of the Manchester, Sheffield, and Lincolnshire Railway, who was called to show that the transit between New Holland and Hull, a distance of three miles, was conveniently managed without a steam ferry proper, by transhipment of passengers and goods from the train to the ferry boats.

The construction of the piers and landing-stages for this service was, as already pointed out, originally made by Mr. Fowler, and it was a high, if not altogether an agreeable, compliment to quote the success of the works at New Holland against their author, when he was advocating a new departure for the improved convenience of a steam ferry proper.

The following letter of condolence from Lord Armstrong (then Sir William Armstrong) was received and preserved by Mr. Fowler :—

"NEWCASTLE, *July* 11, 1872.

"MY DEAR FOWLER,—I am afraid you had an uneasy day of it yesterday. There are days on which all things seem to go wrong, and this seemed to be the case with you yesterday. What on earth could make the Lords' committee declare against your preamble ? It struck me when I was present that Lord Lawrence had some adverse crotchet in his head—something, perhaps, about the supposed impolicy of encouraging a French harbour available as a basis of invasion opposite Dover. However, I can now only wish you success another year."

Mr. Fowler felt this repulse very keenly. Committees of the Houses of Parliament are not obliged to give reasons for their verdict, and they are generally wise enough to avail themselves of this privilege. Undoubtedly the hostility of the two English railways, whose passengers were to be the beneficiaries of the improved Channel service, would have done much to hinder the successful working of the scheme.

Mr. Fowler retained his belief in the practicability of a Channel Ferry to the end of his life. In March, 1882, he contributed a paper to the *Nineteenth Century*, in which, after giving a curious account of the several proposals which had been made for the

P

more convenient crossing of the Channel, he sets out
his own alternative scheme as

"a project for the establishment of huge floating railway
stations which would traverse at a high speed the distance
between the English and French coasts. That is to say, a
Continental train from Victoria or Charing Cross would run
into a first-class station at Dover, and then straight ahead
on to and between the decks of a very large ferry steamer."

He then describes the enlargement of Dover Harbour
and the hydraulic lift, and concludes :—

"Two lines of rails were to have been laid along the lower
deck of the steamer, on which the passenger carriages would
remain in complete shelter, with platforms, waiting and refreshment rooms, and the other conveniences provided in
stations ashore. On arrival in harbour on the French side,
the train would be disembarked by the aid of hydraulic
appliances, and proceed direct on its way, the total saving
of time being estimated at not less than two hours, as
compared with that occupied under the present arrangements. . . . In default of better arguments it has been
attempted to ridicule the system by suggesting that as
passengers would probably leave their carriages during the
Channel crossing, the transit of the carriages would merely
be for the accommodation of the umbrellas and rugs, which
would be their sole occupants. Such an argument hardly
needs refuting, for it would apply equally to the whole
system of through carriage accommodation, which has been
so laboriously built up during the last quarter of a century
in England and abroad. . . . Experienced railway managers
are fully alive to the value of providing through carriages
and all possible conveniences for competitive traffic, and
they know how trifling a matter turns the course of a
traveller along one railway or another. It would, I think,
be difficult to exaggerate the comfort which would result
from the ability to secure a seat at Charing Cross or Victoria

Stations, especially where ladies and invalids are concerned, with the knowledge that there will be no disturbance, no hunting about at Calais or Boulogne in the dark, and no separation of family parties, or necessity to mount into carriages with unknown occupants."

The two railway companies—the London, Chatham, and Dover, and the South Eastern—have now come to terms after a long period of rivalry. As we have already pointed out, the competition which the railway manager has to overcome is the disposition of the Queen's subjects to remain in their own homes. This competition is especially strong when the bourn to which the railway desires to convey its passengers is beyond seas. Now that the harassing and wasteful competition between the two companies has come to an end, we may hope that they will some day find it to their interest to increase the facilities for foreign travel, in which case we may hope to hear again of the Channel Ferry Scheme.

CHAPTER X.

EGYPT

THE year 1869 brings us to an event in Fowler's life which he himself regarded as most important, namely, his first visit to Egypt. On May 8th, 1869, writing to congratulate his father on his birthday, he alludes to his visit of the past winter in the following terms :—

"My own visit to Egypt has been an important incident of the year, and I have been very glad to know you were able to take interest in my wandering and in the sights I witnessed. A thorough exploration of Egypt such as I was able to make must always constitute one of the most important, if not *the* most important event of any man's life.

"The feeling that you are treading on the same ground, that you are on the same wonderful river, and that you are amid the same scenes which in the old days of Abraham and Moses we have always felt to be one of the most interesting pictures of Bible history, must always leave an ineffaceably vivid impression on the mind of those who have been privileged, as I have been, to see the spot itself."

That a man of Fowler's full and varied experience should regard his visit to Egypt—a visit not in the first instance undertaken with any professional object —as perhaps "the most important event of his life," is not a little remarkable; yet the sentiment is, we believe, thoroughly genuine and characteristic of the writer.

The exclusively professional interests of his earlier career have been noticed. By the year 1868 his position as one of the foremost in the ranks of English engineers was assured. Except in regard to some great works, yet to be conceived and executed, his ambition as an engineer was satisfied. Thenceforward he was inclined to allow his exuberant energy and restlessness of intellect to employ itself in other channels as well as in the business of his profession. Unlike many men who have made their own fortunes, Fowler had no inclination to despise the pleasures which find acceptance with an educated leisure class. His faculty for enjoyment was, in this sense, very catholic. He had, moreover, no sort of liking for Bohemian society and Bohemian amusements. The immense amount of thought and money that is devoted to sport in this country is evidence, if evidence is required, that the majority of men who are in a position to indulge their whim, anticipate great enjoyment from its pursuit. Fowler thought so too, and when we come to record his life at Braemore we shall find that even fastidious critics pronounced him to be a good sportsman. So, too, with regard to the company which he kept. Fowler's success in life legitimately gave him an introduction to the best society. He naturally adopted the intellectual independence of tone which characterises the real aristocracy of talent and achievement. He was entirely without what the French call *mauvaise honte*, a social infirmity for which we have no adequate expression. His real kindness of heart furnished him with a courtesy of manner which, though neither courtier-like nor conventional, had the advantage of being obviously sincere. At the same time his conversation was trenchant and forcible. No one could accuse him of being a sycophant. This rugged independence of character was quite compatible with the fact that he made no concealment of his gratification in being honoured with the friendship of the great. He belonged of right to the aristocracy of talent and achievement, and he had a genial pleasure in the position which his abilities and industry had secured for him.

To this catholic yet discriminating choice of interests and recreation, we are disposed to refer the almost enthusiastic interest which he was henceforward to take in the history and destiny of Egypt.

From Fowler's education and other antecedents, one would not naturally have expected that his mental activity would have been inclined to pass much beyond the absorbing engineering problem of the railways and irrigation of Egypt. His mind, however, was responsive to the pleasure found by the scholar and the antiquarian in a study of the ancient history of that wonderful country. We do not mean to imply that he ever became an expert in Egyptian lore, but his attitude towards the scholarly study of antiquity was one distinctly of homage. Occasionally the exponents of modern scientific arts are disposed to throw ridicule on the vocation of the scholar. Fowler had none of this arrogance. Scholarship, a pursuit which had called forth the enthusiasm of so many of the best minds, could not be altogether absurd. He adopted, therefore, in this respect, the verdict of the learned, and endeavoured, with considerable success, to view the question from their standpoint. Egyptology is in scholarship, it might perhaps be said, what salmon fishing, or deer stalking, or yachting is in sport. Fowler was no doubt more assiduous as a sportsman than as an Egyptologist, but his interest in all these pursuits and his love of distinguished society were due to his genial eagerness to enter into those pleasures of life which are approved by the most distinguished and most cultured section of English society.

It was in this spirit in 1869 he first visited Egypt. He had been suffering from the effect of overwork, when, to use his own words,

"a kind and valued friend, the Duke of Sutherland, proposed to me the pleasant remedy of a visit to Egypt with himself and a few friends, including Professor Owen. An expedition to the nearly finished Suez Canal was a part of the programme, and a trip up the Nile under very favourable conditions was suggested as probable. The temptation was beyond my power of resistance, and I gladly agreed to be one of the party."

On board the P. and O. paddle steamer *Nyanza*, on which they embarked at Marseilles, they found the Egyptian statesman, Nubar Pasha, whom Fowler learnt to regard with respect and esteem. The landing at Alexandria was accomplished in state on the "splendid barge rowed by twelve sailors in uniform," which was in attendance for His Excellency Nubar Pasha. In his letters, in a journal kept while in Egypt, and also in a lecture delivered at Tewkesbury in 1880, he describes with considerable minuteness the details of his daily life. Our selection of quotations is compiled in the first place with the object of illustrating the holiday aspect of his journey. We reserve for a later section a brief statement of the professional services which he was invited to render to the Government of the Khedive.

On arrival at Cairo he was carried off to the race-course, where the sand, he remarks, is a poor substitute for the springy turf of Old England. There he was presented to His Highness Ismail Pasha the Khedive, and invited to view the races from his stand. His subsequent relations with Ismail were so important that it seems worth while to preserve Fowler's record of their first interview.

"Ismail Pasha, the son of Ibrahim Pasha, and grandson of Mohammed Ali, was born December 1st, 1830. He is slightly below the middle height, powerfully built, and with very broad shoulders. His habitual expression is somewhat heavy, except when interested during conversation, and then he suddenly flashes a quick look upon you and shows by his manner and by a few emphatic words that he has thoroughly understood all you have said. There can be no doubt of his intelligence, or of his uniformly courteous manner. On these points natives and strangers of all ranks are in accord, however much they may differ with regard to his administration of the country. At the races he seemed greatly amused with the betting, which he encouraged, and when his sons lost their bets to him (which they generally did) he insisted on being instantly paid in gold, and chaffed the young losers most unmercifully. He made rather a large bet with a Captain Butler, who had brought over a mare from Malta, rode her himself, and beat the Khedive's best horse and best jockey. The great Pharaoh was evidently not free from human susceptibilities, for he showed unmistakable signs of annoyance when he lost the race and the bet." "On this occasion," adds Mr. Fowler, "I was introduced by the Khedive to M. de Lesseps."

An early visit was of course made to the Pyramids. After reciting their dimensions and the most probable dates of their construction, he continues :— *

* Tewkesbury Lecture, delivered in 1880.

"As an engineer who has designed and constructed large works, I was naturally much interested in examining the Pyramids, statues, and temples of Egypt, with special reference to their construction, of which we have heard so much during the last twenty years. The first impression on the mind when you come into their immediate presence is that of magnitude. You are positively oppressed with this feeling, especially with the Pyramids of Ghizeh, the Temple of Karnak at Thebes, the statue of Rameses at Memnonium, and the rock temples at Aboo Simbel. Further examination leads you to appreciate and admire the evidences of elaborate and mature design, which are at least as remarkable as the magnitude. Lastly, you find granite, porphyry, alabaster, and other stone of sizes almost unmanageable, which must have been brought great distances by land and water under difficulties nearly insuperable.

"Let us first consider the Pyramids of Ghizeh. The main body of the work is built of blocks of nummulitic limestone quarried on the spot. The two large Pyramids (Cheops and Chephren) were originally faced with fine-grained limestone from the Toorah quarries, of which covering nothing now remains except near the top of the Chephren Pyramid. The third Pyramid of Mycerinus was faced with granite brought from Syene. In the interior of the Great Pyramid are passages, chambers, and sarcophagi formed of polished granite and porphyry, of design and workmanship well worthy of the present time; and in the tombs near the Pyramids are also varieties of granite, porphyry, and alabaster exquisitely worked and polished.

"The ancient Toorah quarries, which furnished the stone for the covering of the Pyramids of Cheops and Chephren, are on the opposite or eastern side of the river, and a few miles further south than the Pyramids. These quarries were really large tunnels cut into the face of the mountain to obtain a particular quality of stone, and extended for several miles in length. The one I explored was entered by an opening about 200 feet wide and 60 feet high, and was half a mile long, with a branch of considerable size for admitting daylight. So distinct are the marks on the rock face, that it is easy to conjecture the size and form of the tools which were used, and the sharpness of the marks indicates, beyond all doubt, that the tools must have been of some very hard metal. The stones cut from these quarries appear to have been of equal size, and of the exact form required to fit the angle of the Pyramid and each other.

"After the stones had been prepared, they were conveyed down and across the river, and from thence up to the Pyramids, on a causeway specially formed for the purpose. They were then fitted with great nicety, without mortar, upon the two Pyramids, polished, and finally inscribed with hieroglyphic writing. These covering stones have long since been conveyed away and used in the buildings of Cairo, but a few fragments have been discovered and preserved.

"The extent and character of these quarries, Pyramids, and temples of a very ancient, if not prehistoric world, being such as I have described, I found it hard to comprehend how persons could bring themselves to suggest the formation of a flat slope of sand round the Pyramids as the means by which they were probably constructed. From numerous hieroglyphic inscriptions we have the certain knowledge that when these works were built Egypt possessed men of science, philosophers, painters, and sculptors, as well as architects, and that the architects who were entrusted with the design and erection of Pyramids and temples were the most noble and distinguished men of the period.

"These facts seem to be quite forgotten, and also the common-sense inference that before such works as the Pyramids of Ghizeh and the Temple of Karnak could have been conceived by a Pharaoh or designed by an architect, earlier and simpler designs necessarily preceded them, and that architects and builders became gradually educated in methods of construction, as well as in principles of design. It is true that we have no record of the metal of which their tools were made, and very little of their constructive appliances, but we have abundant evidence in the works themselves to show that it would be well for us, in these days, if we knew some of the tools they used and how they used them.

"The mutilated statue of Rameses at Memnonium, weighing 900 tons, was undoubtedly brought from the Syene quarries, a distance of about 130 miles, either by a raft on the Nile, or by the less probable way of a prepared causeway for the whole distance; by either plan, the operation would be regarded even now as one of exceeding difficulty.

"A valuable piece of indirect evidence that the removal and elevation of large stones could not in those days have been very difficult to the builders is the fact that they are constantly found in positions where they have no special value either for appearance or structural strength."

Elsewhere, namely in the address given to the Merchant Venturers' School at Bristol, 1893, he says:—

"In Egypt the student of mechanics has abundance of problems on which to exercise his ingenuity. For instance, given the evidence of appliances with which the Egyptians were familiar (as described on the walls at Beni-Hassan and other places), the size of the stones and their position in the Pyramids and temples, especially the statue of Rameses the Second, weighing 888 tons, brought 130 miles from Assouan, then let it be shown in detail how the work was accomplished. A good solution of this problem would be entitled to a first-class prize at the Merchant Venturers' School."

Then, after referring to the discovery of Dr. Flinders Petrie that the tool used for making the statues of hard material (*e.g.* that of Chephren) was probably a precious stone, about as hard as a ruby, set in a frame, he declares his conviction that the works of the ancient Egyptians

"were at least equal to the best works of the present day, as regards (1) Workmanship, as shown in the polished surfaces and the almost invisible mortarless joints of the porphyry, syenite, alabaster, and basalt masonry in the passages and chambers of the Great Pyramid. (2) Statuary, as appears in the statues of hard material. (3) Mechanical Engineering in the removal of vast numbers of huge monolithic masses of rock over great distances and obstacles. (4) Architecture. He must have been a splendid architect indeed who designed the Temple of Karnak in the Plains of Thebes, with its avenue of Sphinxes, its Pylons, its Propylons, its Obelisks, and its Hall of Columns. It has been truly said that the Temple of Karnak was the noblest work ever raised by man for the worship of God."

There is nothing new in Fowler's pronouncement on the engineering skill of the ancient Egyptians, but it is interesting as the carefully worded verdict of an expert who had devoted more of his attention to the question of materials than probably any other engineer of his time.

Fowler described this his first visit to Egypt as a mixture of pleasure-seeking and work. After a few days spent at Cairo he started with the Duke of Sutherland, M. de Lesseps, Professor Owen, and other friends for the Suez Canal, and on his return from his inspection of the great work then on the eve of completion, he was honoured with an invitation to attend the Prince and Princess of Wales on an expedition up the Nile. In this he was accompanied by his guest, Professor Owen, a friend of thirty years' standing. In the course of the trip Fowler was requested to deliver a lecture to the Prince and his suite on his recent visit to the canal, but perhaps the best account we can give of the thoughts which occupied the mind of the convalescent traveller will be by quoting from a letter which he sent to a child friend.

Fowler at all times was a great letter writer. "The well-informed man," he says in his British Association address, "is the man who knows a little about everything and all about something." Another of his axioms was, "Commit what you learn on any subject to writing, it is the best way of remembering and testing your knowledge." Some ideas of this sort may perhaps explain his habit of writing accounts of the scenery, history, geology, and other remarkable but well-known facts connected with the country in which he is travelling. These communications contain exactly the sort of knowledge which would be possessed by the "well-informed."

The following, attuned to the comprehension of a child, seems a favourable specimen of his epistolary style, and supplies a very interesting picture of the practical man of science gossiping with a child on the history of ancient Egypt :—

"On the Nile, Minieh (200 miles above Cairo),
"February 22nd, 1869.

"My dear little Amy,—You will wonder when you see the outside of the envelope of this letter who can possibly have written you a letter with an Egyptian postmark upon it; and when you see the shaky writing, which a shaky, vibrating steamer produces, you will turn to the signature at the end of the letter, and then I hope you will recognise the name of an old friend. And now, my dear Amy, as we are thus introduced to each other, I should like to tell you a little of this wonderful land of Egypt.

"In the first place, let me say that Egypt is remarkable as being the newest, or last-formed land, and yet contains the most ancient people known in the world.

"Let me explain what I mean. In England and France the most recent rocks are the flint-bearing chalk rocks, but in Egypt, long after these chalk rocks were in existence, three

different kinds of limestone came into existence, and rest upon
this chalk, and, as a singular confirmation of the correctness
of this geological order of things, the fossils found in this
Egyptian limestone correspond much more nearly to the existing forms of life on the earth and in the sea, than those found
in the chalk or older rocks.

" Of the antiquity of the people, we have most interesting
records in their temples, their Pyramids, and their now understood hieroglyphics.

" The time of Abraham leaving Haran in the land of Canaan
to take his flocks to drink of the waters of the Nile, and to
escape the famine of his own land by dwelling for a time in
the rich lands of Egypt, cannot be determined with precision
by any reliable historical records, but we know that it was
at an early period of man's history, and that Abraham was the
origin and founder of the Jewish nation.

" In Egypt, however, we have perfectly trustworthy records
of events which happened long before Abraham entered Egypt,
and we have the knowledge that Egypt even then was a great
and powerful nation, with magnificent temples and palaces, and
with her equally magnificent but mysterious Pyramids.

" The earliest records we find in the nature of writing are
the hieroglyphics, which really mean *sacred carving*, and in the
first instance this system was used to represent animals, and
things themselves, but subsequently a kind of language became
formed and understood, and hieroglyphics were then used to
represent the *names* of what was intended to be communicated.

" It is curious that it was reserved for the present century
(about the year 1814) to discover the key by which Egyptian
hieroglyphics could be read, but you will easily understand
how vast a store of new and deeply interesting knowledge was
opened out by it in Egypt, where every temple, obelisk, etc.,
is covered with records which are mysterious now no longer.

" It is almost necessary to come to Egypt and see how
dependent the land is on the water of the Nile, and how good
and bad seasons are produced by the extent of the Nile floods,
to understand how that clever Joseph, when Prime Minister

of Lower Egypt, took advantage of good seasons to change the tenure of the estate and get possessed of the land for the king.

"You remember the land of Goshen, which the Egyptians gave up to the Israelites because it was unfavourably placed for irrigation by the Nile, and required great labour for its cultivation. I thought of this description when I was travelling through the land of Goshen, as it is called to this day, and it is still a true description of it.

"Not far from Goshen I saw a real Philistine from Gaza, which you remember was the scene of Samson's exploits. He was rather a fair-complexioned man, but, like all his race at the present time, an awful scoundrel.

"On crossing the Nile from Cairo to visit the Pyramids you see the place where that kind daughter of Pharaoh found little Moses and saved his life. I always greatly respect Moses, and I hope you do the same. You remember he became possessed of all the knowledge of the Egyptians, and was probably a learned priest of Heliopolis, and then took the lead in forming the new code of laws forbidding the worship of Egyptian gods, and the animals they considered sacred, and enjoined the worship of the one true and living God. Afterwards, when the Israelites had been cruelly used by the Egyptians and their king, Pharaoh, Moses led them out of Egypt, and in their flight from Pharaoh and his hosts they crossed the Red Sea and escaped, whilst the Egyptians, in attempting to follow them, were drowned. Well, my dear Amy, I have seen the place where this crossing is said to have taken place (between the Bitter Lakes and the Red Sea), and I assure you that the rising of the south wind or the tide would account for it by natural laws most perfectly, and we know that God does His work and shows His power by natural laws. I think I have sermonised sufficiently to you, but I assure you it is impossible to forget the Bible when in Egypt. The same ploughs, I believe, are used now as then, and you see the Bedouins of the desert come to buy corn in Egypt as of old, and in numberless ways, places, and things are you reminded of portions of the Bible history of man."

The letter goes on to describe the writer's journey with the Prince and Princess of Wales, an uneventful voyage under the most favourable auspices on the historic river with a distinguished and agreeable company, among whom, besides the staff of the Prince, were Sir Samuel Baker, Mr. Oswald Brierley the artist, the Duke of Sutherland and his son, Professor Owen, and Mr. W. H. Russell of the *Times*.

To Fowler even this pleasure-voyage was not altogether a time of idleness. He was busy in preparing a letter on the Suez Canal, which he sent to the *Times*, and in writing a memorandum on the irrigation of Egypt for the Khedive. This last was presented on his return to Cairo, and Mr. Fowler was requested to delay his return to England in order that the Khedive might have the opportunity of considering and discussing the valuable suggestions therein contained. Announcing this delay to one of his own children, he again dwells on that continuity of history which Egypt seems to suggest to every mind. The famine in Joseph's time, he tells his child, was caused, no doubt, by an inadequate flow of Nile water on the land—

" And it will sound strange when I tell you that my present visit to the Delta of Egypt (which produced the corn which Joseph purchased) is to study the best means of improving the irrigation of the land so as to prevent, as far as the power of man can do, any farther scarcity or famine."

His letter for the *Times*, February 18th, 1869, on the Suez Canal, was transmitted to that journal by the Duke of Argyll, at whose suggestion it seems to have been written. This letter is of some historical importance, as preparing the way for a change of

public opinion in this country towards that great
enterprise.

His inspection of the Canal took place under the
personal guidance of M. de Lesseps, M. Voisin and the
other French engineers, and in the company of the Duke
of Sutherland and Professor Owen.

On the political aspects of the question, Mr. Fowler
was always at pains to defend Lord Palmerston's
opposition to the terms of the original concession made
to M. de Lesseps. This included, he pointed out, (1)
a grant of land along the whole length of the Canal
from the Mediterranean to the Red Sea, with powers
which amounted practically to political independence ;
(2) the absolute property in a fresh water canal from
Cairo, through the land of Goshen, to Ismailia and
Suez ; (3) the supply on the part of the Egyptian
Government of a minimum of 20,000 fellahin on what
was practically a system of forced labour.

These concessions were given to the company
by the Egyptian Government gratuitously; but the
award of the Emperor Napoleon, to whose arbitration certain objections raised by the English and
Egyptian Governments were submitted, obliged the
Egyptian Government to pay a sum of £3,686,000
to the company on its abandonment of the objection-
able part of the concession. Another $3\frac{1}{2}$ millions were
advanced by the Egyptian Government, and with
subsequent concessions and expenditure, first and last,
the canal, Mr. Fowler estimated, cost the Egyptian
Government some 20 millions sterling. The money
had to be raised on very onerous terms. Her
support, in fact, of the great international highway
threw on Egypt an intolerable burden, and, as Mr.

Q

Fowler has remarked on a later occasion, "laid the foundation of the present (1880) large debt and injured credit of Egypt." The benefit to Egypt, moreover, from the construction of the canal was not in any way commensurate with the burden which it imposed.

The party spent five days inspecting the canal. Fowler was busy with the engineering features of the scene, while fossil bones in abundance from the excavations were provided for the amusement of Professor Owen. Among others was a shark's tooth from the head of a prehistoric monster which must, said the Professor, have been 60 feet in length. Nor was the scientific picnic without its mild pleasantries. M. de Lesseps produced, for the inspection of the English savant, a piece of conglomerate in which sea shells were embedded, and demanded a geological identification. The Professor was equal to the occasion, and with confidence assigned the specimen to the *formation Lessepsienne.*

"One day," says Fowler, in his article in the *Minster,* " I had a delightful gallop with Lesseps over five miles of a depression in the desert, now occupied by the Bitter Lakes and forming part of the Suez Canal Navigation, and he playfully reminded me that I should never have another opportunity of a ride on that spot. Lesseps," he says, " took special care to impress upon us that he was not an engineer, and when I requested an interview with the engineers of the canal to give them my impressions of the work, he laughingly declined to be present, as he said he did not understand the elements of engineering, either theoretically or practically."

Fowler's relations with the projector of the great canal were very friendly, and he witnessed with sorrow the failure of his over - sanguine undertaking at the

Isthmus of Panama. Fowler attended the dinner given in his honour at Stafford House by the Duke of Sutherland, on which occasion Mr. Gladstone and Mr. Disraeli were both present to show their respect for the illustrious Frenchman.

"My last interview with Lesseps," Fowler says, "was in Egypt, after the purchase of the Canal shares by England, when he came up to me with both hands extended and said, ' Now, M. Fowler, that your country has become partners with me, and risked money in my canal, England shall share in its management,' and I understand the promise was fulfilled." *

The great canal highroad from Europe to Asia was, as Fowler was fond of pointing out, forced on Egypt by the pressure of Western ideas. The idea of a free waterway worked on commercial principles was one altogether foreign to the Oriental mind. Indeed, the passage and even the approach of foreign ships was a thing to be prohibited rather than encouraged. In this connection Mr. Fowler gives in his Tewkesbury lecture the following translation of a

"Proclamation of the Sublime Porte, under date of the year 1799 (1193 of the Mahommedan). Hatti Chérif,—We do hereby order that no foreign vessel whatever shall approach the Suez Coasts, either openly or secretly. The sea of Suez is, moreover, the privileged route of the glorious pilgrimage to Mecca ; to permit of free navigation to the above-mentioned vessels, to favour and not prevent the same would be to betray our religion, our Sovereign, and the whole of ' Islamism.'

" Consequently whoever dares to transgress this order will inevitably suffer the merited punishment, both in this world and in the next. It is therefore of the utmost importance to the State and to our religion that this peremptory and irrevocable order be conformed to with zeal and promptness. Such is our Imperial Will."

* *Minster*, February, 1895.

Time was when such a decree was not a mere *brutum fulmen*. Much against its will, the Ottoman power has had to abandon its exclusive policy. The very weakness of its rulers as against Western encroachment, coupled with the autocratic power which they exercised over their own subjects, rendered these great concessions to a body of foreign adventurers a possibility. The increasing interest taken by Western powers in the destiny of the far East made them insist on allowing engineering skill to do its utmost to bring East and West nearer together.

England, at the date of Mr. Fowler's lecture (1880), owned 74 per cent. of the shipping that availed itself of the new highway. This fact, and the great stake which she has in her Eastern Empire, combine to make the problem of the internationalisation of the Canal one of vital importance to this country. Here again the art of the engineer has created new problems for the international jurist. Events have proved the incapacity of the Ottoman power to fill adequately the part of trustee to the international waterway which the great Frenchman had pierced through its territory, and have unavoidably thrust that vast responsibility on the country whose interests are most largely at stake.

The development of international law which has arisen out of the opening of the Suez Canal is full of important and hopeful augury for the future history of mankind. Just as the appeal to private warfare has been abandoned for an appeal to the constituted tribunals of the land, so also in international affairs the new values and new properties created by engineering science are brought in case of dispute into newly constituted courts of arbitration, and give rise to new applications of international law.

Fowler's interest in the Suez Canal was merely that of an Englishman and an engineer, but his responsibility for the other engineering problems of Egypt was destined to be very large. The presentation of his report on irrigation led to his appointment, in 1871, to be General Engineering Adviser to the Egyptian Government, an office which he held for eight years.

Apart from the Suez Canal, Egypt is in itself a country created by the engineer, and the political problem which the burden of empire has imposed on this country may be summed up in the sentence that England is under an obligation to give security in Egypt for the work of the engineer. No more striking example of the value of security in the history of progress can be found than that which may be gathered from the success, abandonment, and subsequent resumption under happier auspices of the engineering works on the railways and irrigation of Egypt, instituted in the first place by John Fowler, under the authority of Ismail the Khedive.

The history of modern Egypt, when it comes to be written, will present an extraordinary and interesting contrast to the chronicle of the Pharaohs. A great civilisation is the product of the organisation of labour. That organisation may be directed either by an absolute arbitrary power controlling the movements of vast armies of men, who act not on their own motive, but are marshalled and controlled by the will of a master; or it may proceed from a society in which the parts move on their own free initiative—in other words, from the ordered harmony of economic freedom. The first of these was the basis of the civilisation of ancient Egypt, the civilisation emphatically of a slave power. The Egyptian peasant, whose primitive methods of irrigation Fowler describes, was little further advanced under Ismail than he was under the ancient Pharaohs. Fowler's attempt to bring to his aid the resources of modern science broke down not from any error on the part of the engineer or his methods, but because of the insecurity of the personal despotism of Ismail. The authority of an intelligent Pharaoh, no longer hardening his heart against Western ideas, squandering on one side with Oriental profusion his subjects' money, and on the other promoting schemes of far-reaching and wise beneficence, literally fell to pieces when confronted with the incongruous demand of the bondholder that debts must be paid, and that the government of Egypt must be conducted on business principles. Pharaoh, subjected to the rule of Lombard Street, crumbled away, and with him went alike his waste and his good intentions. It was the fortune of Fowler to be the minister of an irresponsible despot, whose foible was the promotion of beneficent schemes of engineering. He saw the plans which he had laid frustrated by the fall of his patron, and lived to see them resuscitated again under the happier auspices of the *Pax Britannica*.

The *Pax Britannica* is not identical with that popular acquiescence in authority which is the result of free political institutions, but any government that is well administered, though not perhaps the best, still leaves the organisation of society to be ordered by the normal economic principles of personal liberty, security of property, and the right of exchange.

Forced labour, the very basis of the barbaric power of the Pharaohs, has come to an end. Arbitrary spoliation under the title of taxation has been brought within moderate limits. Property is secure and exchange is free. These are the conditions, realised in Egypt under the *Pax Britannica*, which underlie the organisation of the modern industrial cosmos.

The future of India and Egypt, protected by a foreign domination from the ravages of famine, pestilence and civil war, inhabited by peoples whose ambition is confined to the satisfaction of very modest material wants, is likely to produce problems of the deepest interest to the student of history. The question of a surplus population in Western states seems to be solving itself. The rate of increase there is already less rapid, while the growth of wealth is certainly accelerated. Higher standards of life and a keener atmosphere of intellectual activity seem to act as a restraint on the increase of a proletariate population. The increase of wealth, the career open to prudence, and the wider distribution of property have a like restraining influence. In Egypt, and more notably (as the system is of longer standing there) in India, the responsibility of rule, the elevating tasks of intellectual labour, notably the scientific work of engineering, rest for the most part in the hands of aliens. What elements are there in the situation which can prove a substitute for the decimation of the population by war, famine, and disease, those "positive checks," which periodically arose anterior to the arrival of British rule? Better conditions of life in a Western state in any given class are accompanied by a slower growth of population. Increased wealth is used not to produce numbers, but to improve conditions. Among a subject population the same prosperity shows a tendency to produce a different result.

The problem is one which the future only can decide. In Egypt one of the keys of the situation is in the hands of the engineers who are introducing the modern science of engineering into this ancient land, a work in which, under less favourable auspices, Fowler was a pioneer.

It is interesting to record that Mr. Fowler, throughout his Egyptian connection, was at considerable pains to induce native Egyptians to qualify themselves as engineers, and he befriended several young men who came to this country for that purpose. He found, however, an insuperable objection in the mind of the Khedive to give them employment in responsible positions. His Highness preferred to deal with Europeans, remarking to Mr. Fowler, who pressed their claims, that his own subjects of this class assumed for themselves intolerable pretensions.

It is impossible to set out in detail all the work on which Fowler in his capacity of Engineering Adviser to the Khedive was consulted. Three subjects, however, seem to be of the greatest importance : (1) the manufacture of sugar ; (2) the Soudan Railway ; (3) the irrigation of the Delta.

The first was an industry which the Khedive was anxious to promote, more especially on his own estates. At the direction of the Khedive much machinery and plant had been purchased and erected. At Mr. Fowler's suggestion Sir F. J. Bramwell and Dr. Letheby were brought out from England to advise—the one on the machinery and the other on the chemistry of the manufacture.

The growth of sugar on a small scale for home use is not a new industry in Egypt. The Khedive, however, conceived the idea that the industry might be made important, and large new factories had been opened at Minieh, Benisooef, Magaga, Beni Mazar, Matai, and Rhoda. The best machinery was employed, and the result is thus summed up by Fowler in his Tewkesbury lecture :—

"The sugar-cane of Egypt yields a very fair percentage of crystallisable sugar, but is inferior in that respect to the cane of the Mauritius in the proportion of 15 to 18. On the whole, however, it is a question of considerable doubt whether it would not have been more advantageous to the Khedive and to Egypt to use the land for ordinary crops of corn and cotton rather than for the sugar-cane."

Fowler was not responsible for the initiation of these costly and on the whole unprofitable experiments, but much of his time was occupied in the vain endeavour to put the Khedive's speculation on a satisfactory footing. It may be added that the prospects of sugar cultivation in Egypt have since that date considerably improved.

The question of the Egyptian railways, more particularly that of the Soudan Railway, is one of historical as well as personal interest. The facts we shall endeavour to narrate as far as possible in Fowler's own language. In the printed edition of his Tewkesbury lecture he writes as follows :—

"The discoveries in Equatorial Africa during the last twenty years naturally led to the consideration of the means of providing improved communication for the populace and produce of those vast districts. The River Nile is an available navigation, notwithstanding the interruption of the first cataract, as far south as Wady Halfa; beyond that point the constant succession of cataracts make it practically useless, and consequently all goods traffic and personal communication are limited to camel caravan and the electric telegraph.

"The late Khedive instructed me to make the requisite surveys and investigations, so as to advise the Egyptian Government on the important question of providing the best access to these newly discovered regions which had been claimed as a part of the Egyptian territory, and the claim had been unquestioned.[*]

"The result of two years' study was a report to the effect that a light and cheap railway from Wady Halfa to Khartoum (550 miles) would be of inestimable value to trade, government, and civilisation, and that in a few years the traffic would probably he sufficient to give a return on the outlay, but that it would be undesirable to make any railway into Darfour.

"The importance of the railway to Khartoum cannot be over-estimated, either for its immediate or remote influence. It will furnish the cultivated land near the Nile with an outlet for its produce, which at present it does not possess, vast regions on both sides of the railway will have access to it by camel routes, and, at its Khartoum terminus, the Blue Nile for a short distance and the White Nile for a very long distance will continue the communication into Central Africa.

"After due consideration I was authorised to let the works by contract and proceed with the construction of the railway, called for convenience the 'Soudan Railway,' and about 60 miles of its length have been completed and are available for traffic. Unfortunately, financial difficulties have pressed

* Mr. Fowler always insisted that the Soudan was from geographical considerations necessarily a part of Egypt. He wrote in this sense April, 1885, to Earl Granville, urging that no stable government in Egypt was possible which did not include Khartoum and the Soudan. Lord Granville wrote thanking him for his letter, and added, "your authority is very great, but Khartoum appears to me to be a large order."

so heavily upon Egypt that this great and useful work has been interrupted, and will probably for the present be abandoned. My surveys and reports, however, will remain, and when Egypt has recovered from the effects of its wasteful extravagance this work and others on which I have reported—such as the Barrage—will be resumed and carried out.

"I must briefly describe the ceremony of the inauguration of the Soudan Railway, and the speeches delivered on the occasion. On January 15th, 1875, in a violent gale of wind and sand, the ceremony commenced. The Egyptian officers engaged upon the work were in attendance, and soldiers were placed instead of flag posts to mark out the line. Close at hand were the labourers, with their tools and baskets. I verified the levels and angles, and then the Cadi, the ecclesiastical judge, and his assistant with a large assembly arrived, carrying banners, on which were texts from the Koran. The assistant read an address in Arabic, and while he was reading a long piece of poetry incorporated in the address, Chabim Pasha, the Governor, intimated to the orator that sufficient had been recited, upon which the assembled company said 'Amen,' and prematurely ended the address. The Pasha then responded in Arabic."

The history of Fowler's Egyptian employments is the same strange mixture of Western and Oriental ideas. The following, extracted from a report to His Highness, describes the formidable nature of surveys undertaken in the cause of railway construction.

The surveying expedition, which started in November, 1875, and returned to Cairo in August, 1876, contained eight English engineers and a doctor—Messrs. Bakewell, Simpson, Ensor, Chambers, Solymos, Murcott, Burr, Meley, and Dr. Lowe—four Egyptian engineers and officers in command of troops, thirty-six soldiers, 300 to 400 camels, 100 to 120 drivers, guides, and others. In view of the subsequent lapse of these

regions into a state of cruel and fanatical barbarism,
it is noteworthy that the expedition was well received.
In the words of the report, "Every possible facility
was afforded to my surveyors for carrying on their
important work." A curious incident with regard to a
chronometer stolen from this expedition is worth
mention. After the occupation of Omdurman by the
Sirdar in 1898 this chronometer was found in a house
occupied by the Mahdi, as the makers, Messrs.
Frodsham and Co., relate with pardonable pride, in
excellent order and as good as new.

Fowler's diaries for his several Egyptian sojourns
are full of the discussion of large engineering projects.
Interviews with the Khedive were most satisfactory.
Never was there a potentate more desirous of his
subjects' welfare. Distinguished English engineers—
Armstrongs, and Rendels, and Bramwells—flit across
the scene. The railway was of course only part of the
larger question of communications generally. With a
view of facilitating transport of material great works
were instituted at the harbour of Alexandria, but
perhaps the most interesting problem before the
English engineer was the discovery of the best mode
of bringing shipping up the first cataract. The
following, extracted from the diary of 1872, a good
example of how the time was spent, deals with this
point.

On January 27th, 1872, a little before ten o'clock,
Mr. Fowler, accompanied by Sir William Armstrong,
Messrs. Rendel and Knowles, set out from Cairo
to inspect the first cataract. The diary describes
in some detail the carrying of a train across the Nile
at Boulag on a steam pontoon fitted with double rails.

The idea, *mutatis mutandis*, suggests Fowler's larger proposal of a Channel ferry (1872). The pontoon, however, he remarks, is inadequate for the growing traffic, and a railway bridge is, in his judgment, a necessity of the immediate future. The train stopped next at Magaga to allow an inspection of the sugar factory, and finally arrived at Minieh at six o'clock. At Minieh next morning the journey was interrupted by a message from the Khedive that he would receive Mr. Fowler at the Minieh Palace that morning.

Admitted to the presence, Fowler stated that the ferry at Boulag was inadequate; that a bridge was necessary, which would take two years in construction, meantime that the passenger bridge, Kazr-el-Nil, then nearly completed, could be temporarily used for trains; that the railway between Cairo and Alexandria was sufficient, being a double line, and that a rail from Ghizeh to Alexandria was in the meantime unnecessary; that the carrying power of the rail from Rhoda to Ghizeh might with advantage be increased by doubling the line. His Highness then went into the question of the Soudan Railway. At 3 p.m. they started again on the dahabieh *Assouan*, towed by a Government steamer. They anchored for the night at Rhoda. Next day they inspected sugar mills, and continued their journey at twelve o'clock, arriving at Daroot at three, where a locomotive and carriage were waiting to take the party to inspect the new sluice works on the Ibrahimia and Yoosef Canal. Elaborate particulars supplied by the resident Egyptian engineer are entered in the diary. Next morning (January 30th) at six the party started again, and arrived at Sioot, where Mohammed Abusamra, resident engineer for this part

of the Ibrahimia Canal, called and arranged for an
inspection of works next day. The mouth of the
canal, made some four years previously, owing to the
absence of regulating sluices, had been much injured
by floods, and works were then in progress to remedy
this.

On February 1st the steamer arrived at Soohag,
where the local canal had to be inspected. It was dry
and dilapidated. At Thebes the party stopped for a
few days to visit the temples, and on February 6th
they were again ascending the river, and the engineers
were attacking the problem of the cataracts.

"During the earlier part of the day," says the diary,
"Mr. Fowler was engaged with Sir William Armstrong in
considering and discussing details of the hydraulic machinery
proposed to be used for the purpose of working an incline
from the lower to the upper part of the cataracts near
Assouan, and thus improving the navigation of the Nile; also
in calculating the best means of working the river traffic
which will be created with the construction of the Soudan
Railway."

Next day, February 7th, they arrived at Assouan,
and in the afternoon Mr. Fowler took a small boat
as far as the lower part of the cataracts, and landed
near the Isle of Shellal. He then walked along the
east side of the cataracts, and passed the village of
Koroor, returning to the dahabieh by river.

"Next day, February 8th, he was occupied all day with
Sir William Armstrong in examining the various rapids along
the whole course of the cataracts, so as to select the one most
convenient upon which to locate the hydraulic wheels and
pumping station."

The following days were spent in the same manner, and the ground for the proposed inclined slope was surveyed and the gradients laid down. After their return to Cairo, on February 23rd, Mr. Fowler, Sir William Armstrong, and Mr. Rendel had an interview with the Khedive at which Mr. Fowler explained, with plans, the Cataract Incline, and the system by which it was proposed to utilise the water power of the rapids. He pointed out that the scheme suggested would be efficient and economical in working, and that, in point of cost and time required for execution, it was greatly superior to the alternative of a canal, which would involve deep cuttings in the rock, at an expense in money and time difficult to estimate and perhaps prohibitory. Sir William Armstrong showed plans of the water wheels, pumps, accumulator and hauling machinery, and stated that 125 horse power was required for working the water wheels, and that this amount was procurable from one stream in the rapids at all times of the year.

His Highness asked Mr. Fowler whether Nile steamers and large boats could be carried over the incline, and Mr. Fowler replied that all the present Nile vessels could be so carried. His Highness wished to know whether there was space provided for shunting vessels, so as to repair them, if necessary, when hauled out of the water. Mr. Fowler pointed out the area destined for workshops and repairing yard, and further showed the wharf provided by the plan, at which two large and three small vessels could lie. Mr. Fowler also called attention to the harbours of still water at the end of each incline.

As to cost, Mr. Fowler stated that about £50,000

would cover the cost of all the machinery delivered at Alexandria, and that £40,000 would be about the cost of the masonry and other works requiring skilled labour. The bulk of the incline embankment and works could be done by unskilled labour and filled in from material already at hand, and would require about 1,000 men, if the Incline Railway was to be made in one year. Sir William Armstrong agreed that it would take two Niles, or say two years, to put the whole machinery into actual position and operation. The first thing to do would be to build workshops at Mahatta; the machines in these shops would be worked by the water power of the incline, which would also, when not fully engaged in hauling, pump water for irrigation.

His Highness was much struck with the proposal of utilising the power of the rapids, and on learning that at the first cataract it amounts to about 9,000 horse power, at present wholly unused, and that the second cataract was at least as powerful, his sense of economy was shocked. He expressed his regret that such power was not being utilised, and suggested that cotton might be spun as well as grown in Egypt and irrigation much extended, developments of enterprise which, Mr. Fowler said, had already occurred to him.

Mr. Fowler proposed the construction of two trial steamers for the traffic in connection with the incline, which should be so designed and engined as to consume half the coal required by existing steamers. His Highness asked whether existing steamers could not be furnished with improved engines, and proposed that Mr. Fowler should examine the present steamers and select one or two of the most suitable form for the purpose of

being re-engined. He was also desirous that the two
engines should be able to burn wood, which could be
brought down from Wady Halfa at high Niles in large
quantities.

On the subject of coal Mr. Fowler expressed an
opinion that the samples furnished by M. Monnier
from a small shaft at Edfoo were not true coal of
the coal measures, and that such coal would not be
of use in its present condition for engines, though it
might be burnt as fuel of an inferior kind and would
produce gas. At the same time Mr. Fowler desired
to see M. Monnier in concert with Sir William
Armstrong and to report to His Highness after full
inquiry. His Highness stated that the price of coal
must if possible be reduced, and asked Mr. Fowler's
opinion of a plan by which His Highness was himself
to build steamers expressly for coal traffic, and to
import coal, at a loss if necessary, in order to reduce
the present freights. Mr. Fowler replied that it would
be well to use this plan as a menace, and resort to
it only on its failure as a menace. He received His
Highness's authority to confer with coal owners and
shippers in England with the view of procuring purer
coal at low freights. He also pointed out that it
was cheaper to import good coal at a fair price from
England than to import inferior coal at a gift, and
he stated that during his recent journey on the Nile,
he had noted the miserable quality of the coal supplied.

His Highness insisted on the importance of the
question of fuel supply, as bearing on the Soudan
Railway scheme. He said that the Soudan Railway
must be made, but reminded Mr. Fowler that its
success depended in a measure on the cost of steam

R

transport on the Nile. In reply, Mr. Fowler reverted to his proposal of improved marine engines, by which in his opinion a saving of 50 per cent. could be effected in fuel, whether coal or wood.

Mr. Fowler then referred to the necessity of improving the cranes and dock machinery at Alexandria, and obtained leave to confer with Sir William Armstrong for the erection of the necessary plant.

Mr. Fowler next remarked on the Nile deposit above the first cataract, and suggested that, the wind being constant in this district, windmill pumps might be used for irrigation. Nubar Pasha, who was present, said he had tried these without success, but Mr. Fowler pointed out that better results might be obtained in the narrower part of the valley, where the winds were more constant.

A long discussion then followed as to the management of the sugar factories and the private estates of the Khedive generally, in the course of which proposals were made for additional machinery, locomotives, and, a point on which Mr. Fowler was insistent, an organised staff of competent engineers.

It will be convenient to disentangle from other contemporary and pressing matters the history of the railway. A section of the railway was actually completed, but before long the pressure of financial distress brought the works to a standstill. A less ambitious programme became necessary. Accordingly, in January, 1877, we find Mr. Fowler conferring with His Excellency Ismail Pasha Ayoub, Governor of the Soudan, who had undertaken to complete the line by means of the resources of the Soudan. At an interview with the Khedive Mr. Fowler explained to

His Highness that the result of his conference with Ismail Pasha and Mr. Janson was that he was prepared to advise His Highness to proceed with the works of the Soudan Railway, at such a rate of progress as would be consistent with the financial resources of the Soudan, as far as Hanneck, and that from Hanneck to Dabbeh (a distance of 258 kilometres), the river, which was there navigable at all seasons of the year, should for the present be adopted ; that from Dabbeh the railway should be made to Khartoum. On January 19th Mr. Fowler and Mr. Duport had long conferences with Ismail Pasha Ayoub, the Governor of the Soudan, and agreed on a memorandum, which, with only a few alterations, was subsequently accepted by the Khedive :—

"(1) The superior control and responsibility to be vested in the Governor of the Soudan.

(2) Existing financial obligations to be discharged by the Egyptian Government.

(3) After January 1st, 1877, financial responsibility to be on the Soudan Government.

(4) All the net revenue of the Soudan and Equatorial provinces to be devoted to the construction of the railway. Annual sum of at least £50,000 to be paid to Bank of England.

(5) Mr. Janson or other chief engineer to be responsible for direction of works, etc.

(6) Mr. Janson to keep Mr. Fowler, the consulting engineer of the Egyptian Government, advised, and to send lists of material required under the contract with Messrs. Appleby.

(7) The entire management of all matters connected with Soudan Railway to be in the hands of the Governor of Soudan.

(8) To avoid delay under above clause, the Governor of the Soudan authorises the engineers, Mr. Janson and Mr. Duport, Mr. Fowler's responsible representatives in Egypt, to make selections and carry out requisitions.

(9) The transit administration to forward material to Soudan at once, on the certificate of Mr. Duport.

(10) For purchases and material and for salaries the Governor of the Soudan is to provide to the extent of £10,000 per annum.

(11) The telegraph to be under control of Soudan Railway administration.

(12) The Governor of Soudan to submit a yearly financial report to Minister of Finance.

(13) Certificates for material required in the contract with Messrs. Appleby to be signed by Mr. Fowler and to be received by Minister of Finance as a full discharge."

On January 26th, 1877, Mr. Janson saw the Governor of the Soudan, H.E. Ismail Pasha Ayoub, with reference to the railway, apparently for the last time. During the first fortnight of February there is record of numerous interviews with Colonel Gordon, who "expressed himself very strongly and fully on the subject of the cruel and unjust treatment of the natives by Ismail Pasha Ayoub."

On February 14th at half-past eleven Mr. Fowler called at the Palace by appointment, and found His Highness the Khedive, Barrot Bey, Cherif Pasha, Mr. Appleby, and Colonel Gordon. The scheme of organisation for the railway was settled item by item.

"It was explained by His Highness that Gordon Pasha would be the Governor-General of the whole of the Soudan from Assouan to the Equatorial Lakes, and would include the Directorship of the Soudan Railway in his duties.* The alterations made were unimportant, chiefly the adoption of Gordon Pasha as the Governor, and the railway administration under General Marriott for the reception and forwarding of Soudan material."

An appointment was further made for a meeting between Colonel Gordon and Mr. Janson two months hence

"to examine the site of the railway terminus at Hanneck, and also the proposed site of the Kohé Bridge, with a view to determine the question of making a bridge across the river in the first instance, or postponing the bridge and working a ferry provisionally."

Next day, February 15th, Mr. Janson was directed to call on Colonel Gordon with a copy of Messrs. Appleby's contract and the engineers' agreements. Gordon would have nothing to do with the latter, which concerned Mr. Janson and not him. He wished to be troubled as little as possible with the management of the railway, and said he intended to leave the direction of the works entirely to Mr. Janson. He could not at present say what sums he would be able to devote to the construction of the line, but he would give every assistance in his power. He was going to Khartoum as quickly as possible, and would then at once telegraph to Mr. Janson, who, after making all necessary arrangements at Cairo, would then hurry up to meet him at Hanneck and go over the line with him.

* A letter from the Khedive to Gordon, February 17th, 1877, contains the following instruction : "I direct your attention to two points, viz., the Suppression of Slavery and the Improvement of Communications."— BOULGER'S *Life of Gordon*, vol. ii. p. 3.

Gordon Pasha said he would remit to the Bank of
Egypt the sums necessary for payments to Messrs.
Appleby. The sums necessary for local expenses should
be sent to Wady Halfa, where the payments and disbursements would be effected under Mr. Janson's direction. He was anxious that Messrs. Appleby's payments
and deliveries should suffer no delay, and understood
that Messrs. Appleby's position was a very painful and
embarrassing one. He would do all that lay in his
power to prevent any difficulties. He expected to leave
on Saturday morning for Suez.

On the 18th Mr. Fowler called on Gordon Pasha,
and at his desire furnished an estimate of the proportion of Mr. Fowler's personal account due in respect
of the Soudan Railway and Darfour surveys. On the
same day he received from Gordon advice of an order
on the Bank of Egypt for £10,000, to be applied in
payment of Messrs. Appleby.

The following characteristic notes from Gordon are
among Sir John Fowler's papers, and relate to this
period. The first, written on the day of the above-
described interview with the Khedive, bears witness
to the assiduity with which Fowler had pressed the
affairs of the Soudan Railway on the sorely over-
wrought and newly appointed Governor.

"*February* 14*th*, 1877.

"My DEAR MR. FOWLER,—I have received your letter, and
I throw myself on your indulgence to excuse me coming or
indeed discussing the railway question any further. I am
utterly ignorant of this railway for the present, and can form
no ideas on the subject, and will express no opinion on it.
The railway has gone on hitherto without me, and it can
surely do so till I get to Khartoum; I will then give it my

first attention and do all that lies in my power. I have so many more things that need my attention, which are to me more important than the railway is for the moment, that I must attend to them.

"Regretting very much that you are not well, and with my kind regards to Mrs. Fowler,

> "Believe me,
>> "Yours very truly,
>>> "C. G. GORDON."

Two days later the new Governor, on his way to his province, sent the following *amende honorable* for the impatient attitude into which he had been betrayed :—

> "SUEZ, *February* 19*th*, 1877.

"MY DEAR MR. FOWLER,—I was indeed sorry I missed saying good-bye to you and Mrs. Fowler. I wished much to do so, for, owing to worry, etc., I fear I was not as courteous to you as I would have been in my general quiet. I hope you will therefore excuse me in consideration of my position and the hurry I was in.

"I have sent an order to the Bank of Egypt for £2,600.* I wish to clear up all these outstanding claims, for they are, like hang-nails, a great trouble.

"With very kind regards to you and Mrs. Fowler, your son, and Mrs. Ferguson,

> "Believe me,
>> "Yours sincerely,
>>> "C. G. GORDON.

"J. Fowler, Esq."

The more important matters requiring the attention of the new Governor may be easily imagined, but as everything in connection with the heroic defender of Khartoum is of interest we make no apology for

---

* This sum was for salaries due to the Darfour staff.

the following, a copy of which is among Sir John Fowler's papers:—

"CAIRO, *February 17th,* 1877.

"MY DEAR MR. GIBBS,—The firman is signed. H.H. has given me the old province of Equator, the whole of Soudan, and the littoral of Red Sea. No one could be invested with greater power. Finance, etc., are all in my hands, and, in a word, I am astounded at the vast commission.

"It will be my fault now if slavery does not cease, but of course I need time.

"Believe me,
"Yours sincerely,
"C. G. GORDON."

And so Gordon proceeded to his Governorship. When he arrived there and began to count the cost his enthusiasm for the railway, never very warm, began to cool. The following letter, addressed to Mr. Janson, the resident engineer of the Soudan Railway Works, shows his state of mind:—

"IN DESERT, *November 19th,* 1877.

"MY DEAR MR. JANSON,—(Such a pen and such ink!) Camels do well enough for tramway draught. I used one for drawing a cannon in Darfour, and they plough with them in Turkey. Find out all about tramways, the inclines, curves, etc., and any prices of tramway plant; we will not need much. I feel sure that H.H. will be glad of it. It will be a great thing to utilise the Nile; it is by far the simpler mode. A railway is too exotic a plant to flourish in these countries. Do not let Fowler put a spoke in the wheel, viz. stop this work, which he will if he can. Said Pasha had a screw steamer at Debba twenty years ago. If you have spare engine power could you work a wire rope railway or tramway around the rapids? Next winter you must take the levels around the rapids for the tramways by your officers at Wady Halfa. I count on your putting the chain of steamers right up to

Berber in 1878. Try and get Appleby to take (off?) 10 per
cent. If he does not I may have to let it go into the floating
debt list ! "Yours sincerely,

"C. G. GORDON.

"P.S.—I can get sleepers up along the river for the tramways."

On February 4th, 1878, there is entry in Fowler's
diary of the receipt of "a long telegram from Gordon
Pasha explaining why he could not supply more than
£20,000 this year for material." Later in the same
month, at an interview with the Khedive, it was
agreed (February 21st, 1878), with regard to the
Soudan Railway, that the whole question should remain
to be decided on Gordon Pasha's arrival; and so for
the time being the work of constructing the Soudan
Railway came to an end.

In the article in the *Minster*, from which quotation
has already been made, Fowler tells the sequel of the
story as follows :—

"The works of the railway duly proceeded, the whole
country was sanguine, and nearly one hundred miles of the
line were more or less finished. But the dark days were at
hand; the Egyptian Government became subjected to a form
of financial restriction which gradually stopped all payments
to the Soudan Railway, and although Gordon had sent down
contributions from the Soudan, . . . he was finally compelled
to telegraph to me, on March 13th, 1878, that it was impossible
for him to provide funds from the Soudan alone, and in his
capacity of director he was obliged to order the works to
be stopped and the contract cancelled.

"A large sum of money was necessarily wasted in
compensating the contractors,* but this was a small matter

* An arbitration took place between Messrs. Appleby and the Egyptian
Government, in which Mr. Fowler was arbitrator. His award gave some
£78,000 to the contractors.

compared with the consequences which necessarily followed, viz. the official abandonment of the country south of Wady Halfa, with the Egyptian garrisons; the attempted rescue of the garrison by Gordon; and the attempted rescue of Gordon by Lord Wolseley.

"It would be too painful to dwell on these sad events, but I cannot feel that the words of the Khedive Ismail, when I saw him in London, some time after he was deposed, were altogether without justification. 'Ah, M. Fowler, you and I were not such fools, after all. If we had been left alone we should have finished the railway to Khartoum for four millions of money, and established permanent government; but instead of that eleven millions of money have been uselessly spent and many valuable lives sacrificed in a military operation, and the whole of the country south of Wady Halfa abandoned to anarchy and slavery.'"

He concludes with a tribute to Ismail's good qualities, which ought not to be entirely overshadowed by the impatience and extravagance which were the defects of his character.

The ex-Khedive's view as to the strategical importance of a railway hardly requires corroboration, but it is interesting to observe that Lord Salisbury, speaking on May 18th, 1899, on behalf of the Railway Benevolent Institution, dwelt on this aspect of the subject at considerable length :—

"We live," he said, "in a time of many industries, and many industries have only a doubtful and precarious existence. There is many a failing, many a doubtful prospect, but this one thing is certain—that during the last half-century the one industry that has pushed forward beyond any other is the railway industry, and there is no prospect that its power will diminish. I say that from my own point of view with sound conviction, because in the Foreign Office we are particularly employed in considering what influence railway expansion has on the destiny of nations. By a tremendous effort of railway creation we have recently conquered the Soudan. No doubt the Sirdar wielded many weapons, and no weapon less surely than that of his own splendid intelligence and skill; but if you go out of that and ask what material weapons he wielded, I should say he won by the railway, and the railway alone—that railway which he built at the rate of about two miles per day from Korosko, almost now to Khartoum. That railway enabled him to succeed where a far larger force, under great intelligence and with great support, lamentably failed. I can imagine nothing more likely to exult and satisfy the dreams of any railway engineer than to think of what the Sirdar had in his hands. Think of building a railway at the rate of two miles a day, across country where there were no tunnels, where there were hardly any cuttings, and no embankments, and where you had an unlimited command of labour and no difficulties about money, and, above all, where you had the use of the splendid skill of Lieutenant Girouard, a lieutenant of French extraction in Canada, a subject of the Queen, who is now the Railway Commissioner in Egypt, whose wonderful skill enabled him to complete this railway with a rapidity and faultless exactitude that contributed in no small degree to the splendid success which his chief accomplished."

The works projected by Fowler and frustrated at this time by the instability of Oriental rule, have, as all are aware, been resumed under more favourable auspices. A railway to Khartoum is part of Lord Kitchener's plan for the civilisation of Upper Egypt.

The remainder of the letters of Gordon preserved by Mr. Fowler relate mainly to questions of account. The correspondence on this head winds up with the following characteristic touch from Gordon. It should be remembered that Fowler's anxiety about the accounts was not solely on his own behalf. The financial position pressed very heavily on contractors, engineers, and the whole staff of Europeans employed in connection with the railway :—

> "ALEXANDRIA, *January* 10*th*, 1880.
>
> "*To J. Fowler, Esq.*, 2, *Queen Square Place,*
> *Westminster, S.W.*
>
> "MY DEAR MR. FOWLER,—I was tossed out of my place so soon that it was not in my power to say a word about anything. However, having to-day received your letter, January 2nd, I have written the Khedive very strongly, and sent your account, and in my letter I have said, 'Take care not to offend a personage of your status.' It is not my fault, therefore, if you are not paid at once. The best of it is that they owe me a miserable £1,200, which, to spite me, they will put into Floating Debt! It is not the Khedive, but Riaz (the dancer of Abbas Pasha) who does this.
>
> "I hope to see you soon, but *be kind*, and do not ask me to dinner, for I am ill.   "Yours sincerely,
>
>   "C. G. GORDON.
>
> "P.S.—Write yourself to H.H., and say it is a Soudan debt, and H.H. may pay it."

One other letter from Gordon is added, though it does not refer to the Soudan Railway, but to a map which Mr. Fowler had been at great pains to have prepared.

> "114, BEAUFORT STREET, CHELSEA,
>   "*April* 21*st*, 1880.
>
> "MY DEAR MR. FOWLER,—I am very much obliged for your kind present of the map, and your lecture, and for your kind letter. Do not be vexed if I tried to obtain the map in the way of trade, for (with the exception of the Darfour part, which is defective) it is the best map that is published, and I think you ought to let it be purchased. I gave the one you gave me through Watson to the King of the Belgians, who was much pleased with it. I think you might let it

J. Fowler Esqr
2. Queen Square Place
Westminster:
S W.

Alex a/
10. 1. 80.

My dear Mr Fowler.
I was tired out of
any place so much., that it
was not in my power to say
a word about any thing, however
having today received your letter
Jany 2. I have written to
Khedive, very strongly,. and sent
your circular & in my letter,
I have said "take care not"
" to offend a personage of "
" your status" It is only

my fault then if you are not paid at once. The best of it is, that they owe me a miserable 1200, which, in spite me, they will put into floating Debt! it is not the Khedive, but Ray (the dancer of Abbas Pacha) who does this.

I hope to see you soon, but be kind, and do ask me to dinner, for I am ill. Yours sincerely be sold; as I have said, the Darfour part is incorrect, but otherwise it is first rate.

"I will call on you in a few days. You ought, through Rivers Wilson, to push your claim in Egypt. They have enough money to make fresh annexations, and ought to pay their debts of the Soudan. A letter from you to the Khedive and one to R. Wilson would settle it.

"Believe me,

"Yours sincerely,

"C. G. GORDON."

One other of the projects of the Khedive Ismail deserves notice in connection with the improvement of communications in Egypt. He very soon became alive to the fact that the Suez Canal—that great work for which his country had undertaken such heavy burdens—had diverted traffic from the port of Alexandria and carried passengers past Egypt who formerly had travelled through and even halted there, thereby contributing to the prosperity of the land. Egypt, formerly a great terminus, had been practically reduced to the position of a wayside station off the main route between the East and the West.

The Khedive, who, whatever his faults may have been, knew most accurately the necessities and engineering possibilities of his country, conceived the idea of making an alternative Suez Canal *viâ* Cairo and Alexandria. In 1883, on the occasion of some friction between M. de Lesseps and the representatives of British interests, Messrs. Fowler and Baker contributed an article to the *Nineteenth Century* recalling the Khedive's plan, and urged it on the British public as one well worthy of their consideration.

"As ruler of Egypt," they say, "the Khedive, in laying out an alternative Suez Canal, had in his mind the attainment of three great objects : (1) to make Alexandria one of the important ports of the world, and to establish docks for the sea-going vessels at Cairo; (2) to provide an alternative ship canal, by which the traffic would be taken through the heart of the country instead of across an outlying desert; (3) to provide high-level irrigation for the cultivated land of Lower Egypt and means for reclaiming a large area of desert and marsh land, at present of no value to the country."

During his official connection with the Khedive Mr. Fowler had been called into consultation, and was prepared to show how these various objects could be attained. The proposal, as matured and set forth by himself and Sir Benjamin Baker in 1883, was as follows :—

"Referring to the map," the article states, "it will be seen that the proposed canal runs from Alexandria to Suez *viâ* Cairo, a total distance of 240 miles. The Nile divides the canal into two portions, which may be best described separately. At Cairo the level of low water is about 39 feet above sea-level, so there will be a current down the two portions of the canal towards the Mediterranean and the Red Sea respectively. The rate of this current will depend upon the quantity of irrigation water abstracted from the canal, but will always be very moderate. Locks are provided where the canal joins the Nile at Cairo, and basins and docks for the accommodation of shipping. From the basin on the left bank of the river the canal wends its way by straight reaches and easy bends to Alexandria, a total distance of 118 miles. At the 36th, the 66th, and the 85th mile locks are provided, as the fall from the Nile to the sea would otherwise lead to a current of destructive rapidity. An additional lock at the 31st mile will be worked during high Nile. It will be seen from the map that for the last 28 miles of its length the canal runs through Lake Mareotis, and that the interference with cultivated land is minimised.

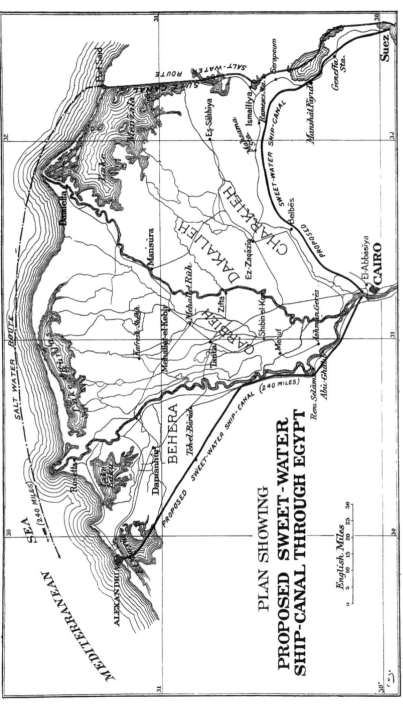

PLAN SHOWING
PROPOSED SWEET-WATER
SHIP-CANAL THROUGH EGYPT

English Miles

To face page 256.

" On the right bank of the river, the canal, leaving the Cairo dock basin, follows approximately the general course of the Ismailia and Sweet Water Canal to Suez, a distance of 122 miles. Locks are provided at the 40th mile and at Suez for use during low Nile, and in flood-time two locks at Cairo and at the 22nd mile respectively would be brought into operation. The works on the canal call for no observation, as they are similar in character to those on the thousands of miles of canal already constructed in Egypt."

The crossing of the Nile by ships in transit from Suez to Alexandria was to be facilitated by "regulation works of some magnitude."

" The spot selected for the crossing is a little below Cairo, where the two flood channels unite and form a single stream of fairly regular and equal flow. Training walls and banks are designed to confine the river to the permanent course for a certain distance above and below the point of crossing, and direct the scour so as to maintain the required depth for the passage of vessels. A railway bridge is provided to connect the lines on opposite sides of the river, and to serve as a carrier for the traversing mooring to which ships would be attached when crossing the Nile. The mode of procedure would be as follows: A vessel, say from Suez, on arriving in the Cairo basin, would be slewed round and passed through the lock into the Nile stern foremost, with her bows pointing straight up stream. A wire hawser would be attached to the traversing mooring, and the latter would be hauled across the railway bridge by fixed hydraulic engines at the Cairo Docks, taking with it the attached vessel. On arriving at the opposite bank the vessel would be in a position to enter the lock to the Alexandria Canal bow first. There is no cross current tending to embarrass the operations, since the locks point sharply down stream, and the ships leave the locks stern foremost and enter them bow foremost. . . . From the experience gained in the working of the great

s

railway ferry near the same spot, it is estimated that ten minutes will amply suffice for the warping of a vessel across the river from one canal to the other."

Such a canal would, it is further urged, be most valuable for irrigation purposes. The portion from Cairo to Alexandria would supply water

" to irrigate without pumping the half - million feddans of cultivated and cultivable land in the province of Behera," and the Suez portion of the canal would meet the irrigation requirements " of about a million feddans in the provinces of Charkieh and Dakalieh, and considerably increase the facilities for irrigating a further area of a quarter of a million feddans in Gallioub."

The expenditure required to construct this ship canal, and to complete the barrage with the required minor irrigation channels, is estimated at from 10 to 12 millions. The water rates alone would, in the judgment of the engineers, give a handsome return independent of the ship dues. The fresh water would cleanse the ships' bottoms from marine growths, and the route would be popular with travellers. It would be at least a day longer than the Suez Canal route, but the dues would be less.

" In conclusion," the engineers put it, " it may be said that the question of the construction of an alternative Suez Canal *via* Alexandria and Cairo resolves itself into this: Is it, in a rainless country like Egypt, preferable to construct a sweet-water canal running along a ridge, or to widen a salt-water " ditch " lying down in a hollow ? and is it, as regards our own country, preferable to have an alternative route for ships through Egypt remote from the present one and under our own control, or to be wholly dependent upon M. de Lesseps and his successors ?"

About the same time a series of articles appeared in the *Times* setting out the desirability and practicability of having an alternative canal parallel to that controlled by M. de Lesseps. These articles succeeded in achieving one of the objects of controversy. They made the adversary very angry.

Happily these differences have been composed, and in the meantime we seem to be working smoothly with the successors of M. de Lesseps ; but a project recommended by such a combination of practical and scientific authority as that of the two engineers whose names are attached to the above - quoted article is well worthy of the attention of administrators and capitalists.

To return to the other important subject on which Fowler was consulted, namely, to the question of irrigation. The following popular description of Egypt taken from his Tewkesbury lecture will give a good idea of the scene of these labours.

" Egypt," he says, " an exceptional country in every respect, cannot be described in the usual manner by definite numbers of miles in length and breadth. A homely illustration is often better than a learned disquisition or an army of figures, so I will ask you to imagine a large kite, with a long string or tail, laid flat on a sandy beach, with its head washed by the sea, and its tail extended straggling inland over the sand. This will give you a good general idea of the form of cultivable land of Egypt and the surrounding desert. As regards size, you must suppose the head of the kite to be a triangle or delta with equal sides of 150 miles, and the tail to be about 2,500 miles long, or nearly as far as from England to America, for that is the length of the Nile up to the Equatorial boundary of Egypt. The area of the body of the kite, which represents the cultivated lands of the Delta and Lower Egypt, is rather more than 2,500,000 acres, or, say, three times the area of Gloucestershire. The string, or tail, represents the narrow ribbon of cultivated land on each side of the Nile in Lower, Middle, or Upper Egypt, which in area is about 2,000,000 acres, or two and a half times the size of Gloucestershire. The area of the whole is rather less than Wales, and the population, without Kordofan and Darfour, about 6,000,000, or the same as Ireland and 50 per cent. more than London. Thus you see how insignificant in size and population is Egypt Proper compared with the interest it excites. . . . If there were no rain in Gloucestershire, the only cultivated lands would be those subject to Severn floods, and such other patches of ground as could be irrigated by pumping water from the Severn. It is thus in Egypt; for in the words of Herodotus, 'Egypt is the gift of the Nile,' and except for the fertilising mud in the waters of the Nile, the whole country from the junction of the White and Blue Niles at Khartoum to the sea would be one vast desert without a single green cultivated spot. The amount of mud brought down by the Nile each year, according to my investigations, is about 150,000,000 tons or more than the whole excavations of the Suez Canal.

" So much for the mud, now for the water.

" The discharge of the Severn at Diglis Weir, during an excessive flood, is 500 tons per second ; that of the Nile at high flood is 10,000 tons per second. I have had occasion to study both rivers very carefully, and I am able to tell you that, on the average, the Severn discharges 3,500 million tons of water per annum into the Bristol Channel, and the Nile no less than thirty times that quantity into the Mediterranean, or 105,000 millions of tons, sufficient to flood the whole area of England to a depth of three feet.

" The value of this vast supply of water to the thirsty soil under the hot sun and rainless sky of Egypt cannot be over-estimated. When the first Napoleon was in Egypt, he declared his determination, if he remained in the country, to use every drop of Nile water on the land, and to permit no discharge into the Mediterranean. He was quite right in the principle which his words indicated, though he could not have literally accomplished it.

"The most ancient work in Egypt for storing the waters of the Nile was the famous Lake Moeris, situated in the southeast part of the land called the Fayoum, which is practically a rich oasis with one side joining the Nile Valley and the other abutting on the Lake Birket-el-karon, whose waters are nearly 100 feet below the level of the Mediterranean. During inundation the waters of the river entered into Lake Moeris by means of a canal, and were retained by locks. At low water, the gates were opened to irrigate the great plains of the district in the neighbourhood of the lake."

In the same lecture he describes the primitive methods of irrigation which it was part of his business to supersede by more scientific methods :—

"First. The *Natala*, the least used, and only for raising water to a small height, is worked by two men swinging a basket by ropes.

"Second. The *Shadoof* is formed of two posts, about 6 feet high and 4 feet apart, with a palm-tree pole placed horizontally across the top; to this a lever is secured by ropes, having at one end a balance weight and at the other a leathern bucket, attached by a long palm stick. The bucket is pulled down to the water and filled, and the balance weight then lifts it to the required height. If the river bank is high two or more of these are used, one above the other, to raise the water to the level of the land. The usual height of each lift is from 8 to 9 feet, and each man raises about 25 gallons of water per minute. The hieroglyphics at Thebes represent the ancient Egyptians working shadoofs of precisely the same construction as those employed at the present day.

"Third. The *Sakieh*, or Persian wheel, consists of a vertical wheel of about 20 feet diameter, and descending to the level of the water is an endless rope ladder with earthenware jars attached. The jars fill as they dip into the water, are drawn up full, and when at their maximum elevation discharge into a

trough or channel from which smaller channels lead into the
lands for irrigation. The sakieh is worked by means of oxen,
or other cattle, attached to a horizontal wheel, which works by
cogs into the vertical wheel described."

A larger measure of irrigation had been inaugurated
on native initiation. This was the Grand Barrage of
the Nile, projected by Mehemet Ali, and completed
(so far as, before the completion carried out by Sir
Colin Scott Moncrieff, it can be said to have been
completed) by his successor Said Pasha.

"The object of this work was to back up the water of the
Nile to a certain level by means of weirs and sluices, so as
to supply irrigation water to the Delta without pumping. The
conception was a grand one, and the principle sound, but
the foundations and other details were insufficient, and the
work failed to accomplish the object of the designer. . . .
When the sluices fitted in the arched opening were first
closed, the water under pressure found its way beneath the
foundations, and, carrying sand with it, undermined and
endangered the whole structure. Practically, therefore, the
Barrage has been of comparatively little use, except as a bridge
for a roadway connecting the opposite banks of the river
with the Delta."

In the diary for 1872 there is the following note
on the Barrage made after an inspection in the
company of Sir William Armstrong and Mr. (now
Lord) Rendel, who were then with Mr. Fowler at
Cairo :—

"The Barrage is situated two miles below this, at the
southern point or apex of the Delta, where the Nile divides
itself into two branches, Rosetta and Damietta. This work
was commenced twenty-five years ago, and consists of a bridge
about 530 metres long, having seventy-one openings, over the
Damietta branch; and of a bridge about 465 metres long,

having sixty-one openings, over the Rosetta branch. The width of the openings are all the same, being 5 metres wide. The thickness of the pier 2 metres each. The width of the roadway of the bridge between the parapets is 10 metres, while the piers are 14 metres in length, that is, the piers extend up stream 5 metres beyond the southern face of the archways ; this extension of the piers forms the point of attachment of the radial arms of the segmental sluice gates. The level of the springing of the arches is 8 metres above low water, the roadway being 12 metres above the same level. The level of the top of the sluices when shut is 4 metres above low water. 'The radier,' or concrete foundation, is 32 metres wide and 3 metres thick, is enclosed between two rows of piling, and extends the whole length of the bridge. The piles are stated to be driven to a depth of from 8 to 9 metres below low water. The level of the upper surface of the radier is 1·80 metre below low water. About three or four years ago, when the sluice gates were closed so as to maintain a head of water of about 3 metres, it was observed that two or three of the arches near the left bank of the Rosetta branch slid in a down stream direction about 6 inches, and settled to some extent at the same time, producing serious cracks in the masonry ; and these arches have since then continued to move every year to a slight extent. No attempt has since then been made to maintain a head of water greater than 0·80 metre. Works are now in progress to ascertain the extent of damage sustained by the foundations, and the means that can best be taken for their repair. Provision is made for the river navigation by means of two locks of 12 and 15 metres wide respectively.

" Between these two bridges, which form a barrage, or dam, across the branches of the river, and at the apex of the Delta, is placed the entrance of a canal, which is 60 metres wide at bottom, 7½ metres deep, with slopes of 3 to 1. It was made for the purpose of irrigating a large area of the Delta, but in consequence of the failure of the barrage, by which the required head of water cannot be attained, this object has been but very partially fulfilled. The barrage over the Damietta branch remains at present unprovided with permanent sluice gates."

"The completion of the barrage," says Mr. Fowler, in his Tewkesbury lecture, "was one of the works which the late Khedive most desired to have carried out. By his instructions I made minute surveys and studies, including observations to obtain the minimum flow of the Nile, and also made borings to a considerable depth near the site of the barrage. These borings penetrated through the Nile deposit into the original marine sand and gravel, and were of considerable interest beyond the object for which they were taken. Eventually I drew up an exhaustive report, design, and estimate for the whole work of rectification and completion, which proved beyond doubt the practicability of the work, and its value in largely increasing the revenue of Egypt at a comparatively small cost. For the sum of about £1,500,000, the Nile floods would be regulated, pumping dispensed with, serious damage to crops rendered impossible, and about 600,000 additional acres of land brought under cultivation. The design was approved, and its execution would certainly have produced a clear profit of more than £1,000,000 per annum, but unfortunately some delay took place and now financial difficulties have postponed it *sine die*."

The great difficulty of the engineering operation was occasioned by the enormous weight of water which had to be upheld, and the terrific force of the rush of water over a vertical fall of several metres. A plan had been submitted to the Khedive by Colonel Rundall which proposed to rely on the existing edifice and to protect the bed of the river by some species of paving. This Fowler condemned as altogether inadequate. His own proposal recommended the erection of a supplementary curtain or wall, which should have the effect of relieving some of the stress thrown on the barrage. Mr. Gatget, a contractor largely interested in Egyptian operations, had made a similar suggestion and had proposed to build the wall or curtain above the barrage. Mr. Fowler's design, however, was to place the curtain wall below the barrage. This device would provide what Fowler calls a "water cushion" which would effectually break the force of the falling waters.

Like the Soudan Railway, the work of Egyptian irrigation has again been renewed. The barrage has been restored by Sir Colin Scott Moncrieff and a staff of engineers who had gained their experience of irrigation works in India; and at the present time Sir John Fowler's partner, Sir Benjamin Baker, is consulting engineer to the Egyptian Government, which is carrying out great supplementary works by making dams across the Nile above and below Cairo so as to control more completely the fertilising flow of the mighty river, and a huge reservoir above Assouan to store flood water for summer use.

All these important works, as has been already indicated, were conducted during a period of financial pressure. The diaries are full of indications of the coming collapse. Under date February 2nd, 1877, the situation is summed up with telegraphic brevity in the following despatch sent by Messrs. Stephenson, Clark and Co. to Mr. Fowler: "When do you expect to get the money from the Khedive?" No satisfactory reply was forthcoming to this and similar inquiries. Barrot Bey, interrogated by Mr. Baldry and Mr. Arthur Fowler, referred them to the Minister of Finance and the Minister of Finance referred them to the Khedive. Messrs. Siemens and Co., according to an entry of the 5th, refused to send further shipments till payment was made. On March 4th Mr. Fowler had an interview with His Highness, when, after suitable apology, he opened up the question of accounts. His Highness admitted the urgency of the situation, and very graciously promised to mention the subject to the Minister of Finance. Many of these sums were personal debts of the Khedive, but all alike they seemed to drift into the "floating debt" about which Gordon wrote with such grim humour. His Highness Prince Hussein, the Minister of Finance, was next seen ; he replied that, knowing Mr. Fowler so well and so long, he desired to effect a settlement as quickly as possible. " He would confer with the Controller, without whom he could make no decision." And so the evasive game went on.

Other negotiations with this courteous but impecunious potentate were more successful.

The diary records how Mr. Fowler co-operated with the British Consul-General, Mr. (afterwards Lord) Vivian, in obtaining the Khedive's leave to remove Cleopatra's Needle to England. The obelisk, packed in a cylindrical sea-going vessel designed by Mr. Benjamin Baker, was brought to England, and erected on the Thames Embankment by Mr. Dixon, the engineer and contractor. The cost, much increased by salvage charges arising out of a temporary abandonment of the vessel in the Bay of Biscay, exceeded by about £7,000 the contract price of £10,000 subscribed by Mr. (afterwards Sir) Erasmus Wilson. The following, from a letter dated Cairo, March 10th, 1877, addressed by Mr. Fowler to Mr. Dixon, is of sufficient interest to warrant reproduction :—

" You and all friends in England will, I am sure, be gratified with the manner, as well as with the fact, of the consent of the Khedive respecting the removal of the obelisk to England.

" Mr. Vivian sends by post to-day the letter of Cherif Pasha giving the consent of the Khedive, and his reply of thanks for the consent and terms of it.

" These letters are so admirable that in my opinion they should be published, which the *Times* will be glad to do, in appropriate type and position.

" As to the best mode of obtaining the publicity, I would suggest that some member of the House of Commons should be requested to ask the Government if there be any objection to their production of the correspondence respecting the removal of Cleopatra's Needle.

" You can easily find a member who will willingly do this, but, before doing so, see Lord Tenterden.

" I saw His Highness the Khedive this morning, and told him of the telegram from London requesting me to thank him for his very courteous assent.

" I must now tell you how immensely we are all indebted to Mr. Vivian, Her Majesty's Consul-General, who has not only given his influential assistance to bring about the decisive result in the kindest possible manner, but has guided all parties as to the proper manner of doing it, and has succeeded perfectly. . . .

" I am glad to be able to congratulate you on the success which has thus far rewarded your perseverance.

" I shall be in London in a short time, and you can always consult me on engineering points."

Next year, 1878, Fowler, shortly after his arrival in Egypt, had an interview with the Khedive, with regard to which there is the following entry :—

" *January* 20.—After some remarks from His Highness as to the financial advantages that would be obtained by the re-construction of the barrage, Mr. Fowler said he told all his friends who were interested in the finances of Egypt, that, in any financial scheme for this country, proper provision should be made for the improvement of the country and for public works, such as the barrage, canals, and the Soudan Railway. His Highness said Mr. Fowler was quite right, and this would be the only proper way of dealing with the question; the Soudan Railway was a most important work with reference especially to the Soudan and the Equatorial provinces, where there were at present from ten to twelve million people shut out from communication with the civilised world. The railway would be the means of opening up and civilising the people, and developing the commerce and industry of those regions.

"Later in the day Mr. Fowler showed His Highness a model of the barrage, when, on being told that the cost would be one and a half million, His Highness remarked that this work would be the best guarantee the bondholders could have. Mr. Fowler agreed, and His Highness added that it would be a better guarantee than all the controllers and all the Caisse. The new Administration had already cost the country in one year £150,000, or about 10 per cent. on one and a half million."

On February 5th Mr. Fowler records that he had a long talk with Mr. Vivian, who said that affairs were in a most critical state, as open war had broken out between the Khedive and the bondholders, who were disposed to make a point of the Khedive's removal before entering into any further *pourparlers* for a new arrangement.

At an interview with His Highness, February 19th, Mr. Fowler explained

" that he proposed to return to England on the following Saturday. His Highness said he wished to speak with him on the subject of the barrage, and asked him whether he had considered means for obtaining funds for that work. Mr. Fowler replied that he had thought of this. He was of opinion that provision should be made for this work upon the occasion of the reconsideration and rearrangement of the finances.

"His Highness said he considered the matter did not concern the bondholders, and that they ought not to be consulted in the matter. The proprietors of the land were the persons interested, and the money required should be obtained on their security."

At a further interview, two days later, February 21st, 1878, His Highness again dilated on the importance of the barrage. He explained that two million feddan then useless might be brought under cultivation, and that the revenue might be immensely increased in a short period. The barrage works, however, must be abandoned till some definite financial arrangement had been made. His intention was that, as soon as the financial difficulties were overcome, the barrage should receive his first consideration, and be carried out with Mr. Fowler's assistance.

Mr. Fowler pointed out that the existing financial scheme was most unfavourable, not only for Egypt but also for the bondholders; that it made no provision for payment of the just claims of the Administration, and none for the maintenance of the existing public works upon which the welfare of the country depends, nor for the construction of the new ones of which the country stands in need. His Highness said that Mr. Fowler was quite right, that he himself had only the interest of the country at heart, that he had asked the British Government to appoint a suitable person as Inspector-General of Lower Egypt, to whom he was prepared to give full powers. This proved that he was acting honestly, although many people were pleased to accuse him of bad intentions and of hiding something. This invitation to the British Government was sufficient answer to

such accusations, and he only regretted that they had
refused to accede to his request.   He hoped they
would reconsider their decision.   Mr. Goschen, he said,
was too much occupied with his own personal political
position at home ; he had adopted his scheme hastily
and without due consideration, and it was an evident
failure, and he was now endeavouring by calumny
and all means in his power to throw the fault of
that failure on His Highness.

Mr. Fowler pointed out that it was a very serious
thing for anyone in Mr. Goschen's position, and with
his financial reputation, to acknowledge the failure.
His Highness replied that the failure was evident,
and must be understood by everyone, but that Mr.
Goschen was a very obstinate man and was endeavouring to cover his own failure by laying blame on His
Highness.

Mr. Fowler said he had known of His Highness'
request to the British Government to appoint an
Inspector-General, and in an interview he had with
Lord Derby, whom he knew very well, he had
recommended that if the Government did send anyone in that capacity it should be a man of high
position, so that his advice might be of real service
to His Highness.

On February 23rd there is further mention of the
financial situation and of an interview with Mr.
Vivian.

"With regard to Mr. Goschen's scheme, Mr. Vivian agreed
with Mr. Fowler that it was an incomplete scheme. Neither
Mr. Vivian nor Mr. Fowler thought that the payment of the
next coupon when it fell due was possible, unless there was
money in the State chest unknown to the Controllers."

Mr. Fowler called afterwards on Barrot Bey, with whom he had a general conversation. Mr. Fowler expressed his opinion that even if a full and complete arrangement was made to meet all debts (as well as the bondholders) the country would even then be able to bear the strain upon it, provided that the finances were ably and skilfully handled.

To render clear the above expression of opinion, a brief summary of the history of Egyptian finance is required.

"The first general settlement of all liabilities, which was effected on the proposals of Mr. Goschen and M. Joubert in November, 1876, was of very short duration. The arrangement, indeed, was a reasonable one, upon the facts as stated to these gentlemen, but unfortunately the statement was wholly misleading." *

On the breakdown of the Goschen arrangement some new scheme had to be proposed. It was the view of the Khedive, of Fowler, and of the illustrious Gordon that the situation did not entirely belong to the bondholders. The opinion of the Khedive was not perhaps entitled to much consideration. The debt was his autocratic creation, and his personal responsibility to the bondholders was undoubted. Fowler was of opinion that, with proper management, the bondholders might be paid, and that provision might still be forthcoming for the engineering development of Egypt; but that in any case the development of the country should not be postponed to the rights of the bondholders. Gordon's position was not very different, though naturally the expense of establishing good government as conceived by him included less engineering than

* MILNER, *England in Egypt*, 2nd edition, p. 220.

in the estimate of Fowler. The Khedive, worried to
death by the demands of the representatives of the
bondholders — the financial cormorants, as Gordon
called them—in January, 1878, summoned Gordon to
Cairo from the Soudan to help him over his financial
difficulties. Gordon, in March of this year, agreed to
accept the post of president of a Commission of
Inquiry. With him was associated M. de Lesseps,
and the understanding was that the Commissioners of
the Debt, who were regarded as the representatives
of the bondholders, should not be on the Commission
of Inquiry. The story of this extraordinary transaction is told at length in Mr. Boulger's *Life of
Gordon*. Lesseps took little interest in the matter,
and surrendered to the pressure of the French Government, which, with the other Great Powers, insisted on
the Debt Commissioners being placed on the inquiry.
The representatives of all the European Powers were
against Gordon and the Khedive. Gordon desired
to have money for the payment of salaries in arrear
and the redress of grievances, and even proposed to
obtain it by repudiation of the next coupon. "Ismail
was not equal to the occasion. He shut himself up
in his harem for two days, and, as Gordon said, 'the
game was lost.' "*

The facts and the equity of the situation are very
much involved. With Gordon and with Fowler we
may doubt the right of a spendthrift autocrat to
pledge the credit of his impoverished serfs for payment of his debts. It is represented, however, and
forcibly enough by Milner, that the burden from which
Egypt was suffering was not the debt to the bond-

* BOULGER, vol. ii. p. 23.

holders, but the incorrigible dishonesty of poor Ismail himself. The first thing to be repudiated was the ruler. This, as we all know, was done, but the difficulty remained that there was nothing satisfactory to put in his place. Ismail was, without doubt, the ablest of his family. Even Gordon, who knew how little he was to be trusted, regretted his abdication. "It grieves me," he writes, "to think what sufferings my poor Khedive Ismail has had to go through."*

Subsequent events seem to show that there was justice in both views. When Ismail was removed, after an interval during which progress was much obstructed by the jealousies of the dual control, the finances of Egypt under skilful management are proving equal to the discharge of Ismail's debts and to the gradual development of Egyptian resources by works which in that country seem necessarily left in the hands of the Government.

In 1879 Mr. Fowler returned to Egypt at the end of January. The diary tells of conferences with Mr. Rivers Wilson, M. de Lesseps, and with the Khedive, with respect to his engineering undertakings. The end, however, was fast approaching, and no satisfactory arrangement could be made for the continuance of the works.

Tuesday, February 18th, 1879, is marked as a

"famous day, from the 'demonstration' by Egyptian officers to demand their pay at the Ministry of Finance, leading to a slight assault on Nubar Pasha and Rivers Wilson, and the appearance of the Khedive on the scene with troops, who quelled the affair after a few persons had been wounded. Nubar Pasha resigned next day. On the following days no business was done, owing to the confusion and excitement."

* *Story of Chinese Gordon*, HAKE, vol. ii. p. 358.

T

Mr. Fowler enters in the note that he called on
Rivers Wilson and begged him not to resign, as he
would sacrifice himself and do no good by such a step.
On Wednesday, March 13th, Mr. Fowler wrote to
the Khedive a farewell letter. On March 15th he had
a farewell interview with the Khedive, at which that
courteous potentate made a great many handsome
speeches. And so John Fowler passed out of the
tangled skein of Egyptian politics.

The surveys of the Soudan and much other information collected by Mr. Fowler were subsequently placed
at the disposal of Her Majesty's Government.

The *London Gazette* of Tuesday, September 1st,
1885, contains the following announcement :—

"*Chancery of the Order of St. Michael and St. George,
Downing Street, September 1st.*

"The Queen has been graciously pleased to give directions
for the following appointment to the Most Distinguished Order
of St. Michael and St. George :—

"To be an Ordinary Member of the Second Class, or
Knights Commanders of the said Most Distinguished Order,
John Fowler, Esq., c.e., for services rendered to Her Majesty's
Government in connection with the recent operations in Egypt
and the Soudan."

A question addressed by his son, the Rev. Montague
Fowler, to the Commander-in-Chief as to the precise
nature of Sir John's services has elicited the following
interesting reply :—

"Commander-in-Chief, *January* 12*th*, 1900.

"Dear Mr. Fowler,—In 1884 Sir John Fowler most kindly
placed in my hands copies of the surveys he had made for a
railway from Assuan to Khartoum.*

* For Fowler's opinion as to the superiority of the Nile route over the
Suakin Berber route, see p. 342.

"They were invaluable to me in 1884-5, and helped the army very materially in our advance upon Khartoum. They were well executed, and enabled my operating columns to find water where no other available maps informed me it was to be had.

"I feel I owe your father a deep debt of gratitude for all the information he furnished me during that too lately undertaken expedition to try and save General Charles Gordon's life. You are at liberty to make any use you wish of what I say here.

"Believe me to be,

"Very faithfully yours,

"WOLSELEY."

To complete the list of Sir John Fowler's engagements outside the United Kingdom, two later expeditions should here be mentioned. In 1886, accompanied by his wife, he went to Australia, and inspected the railways of New South Wales, of which for forty years his brother-in-law, Mr. Whitton, had been engineer-in-chief. Sir John remained consulting engineer to the Government of the Colony till his death.

In 1890 he visited India, taking with him his eldest son. Here he was hospitably entertained by the Viceroy, Lord Lansdowne, and by the governors of Bombay, Bengal, and Madras, all of them old personal friends. A special train was put at his disposal during the period of his visit. He was consulted by Lord Roberts on the frontier railways, and by the Government on a variety of engineering problems. This flattering reception was a fitting recognition of a long career of industry and usefulness.

CHAPTER XI.

THE FORTH BRIDGE

WE now come to the last great work with which Fowler was connected, the crowning victory of engineering science in the nineteenth century, the bridging of the tidal estuary of the Forth.

A reference to the map of Scotland will show that if .a line is drawn from Dunbar, on the Haddington coast, to Anstruther, on the opposite shore of Fife, there will be to the west of it a firth of some 62 miles long. Till comparatively recent years there was no bridge across this great flow of water till we reached its westernmost point at Stirling. The Lothians, Fife, and Kinross contain some of the richest mineral and agricultural districts in the United Kingdom, but up to the year 1885, when the Alloa Bridge was made, all traffic between the north and south of the Forth had to go round by Stirling, or be carried over the firth in ferry-boats. The bridge at Alloa was some 42 miles to the west of the imaginary line above described. There were three principal ferries across the water. That best known to passengers of the present generation, from Granton to Burntisland, which was twenty-four miles up the Forth ; next, some eight miles higher up the river, the passage familiar to readers of *The Antiquary*, from South to North

Queensferry, close to the site of the great bridge; and a smaller and less important crossing, to Kincardine, some fifteen miles further to the west. In addition to the great inconvenience caused to local travellers and traffic, a vast amount of goods in transit to the northern part of Scotland was brought into Edinburgh by the great trunk lines from the south, and there delayed. The railway passage at Stirling and at Alloa was in the hands of one company, which naturally was not inclined to give the fullest facilities to its competitors.

As we noticed in discussing the Channel Ferry scheme, there was between Granton and Burntisland a service of ferry-boats, which carried goods trains, or portions of goods trains, from shore to shore, but the transit of passengers was attended with the most terrible discomfort and inconvenience. After a short railway journey from Edinburgh, the passengers with their luggage were turned out at Granton at a miserable station. Laden with wraps and hand-packages, they had to stagger down over rough cobble stones to the steamer. Here, at a narrow gangway, a stern official demanded the ticket of the exhausted traveller, who had to drop his packages in the wet, and search his pockets for the missing bit of cardboard. The heavy baggage was brought down in hand-barrows by perspiring porters. The descent, when the tide was low, was steep, and the only brake employed on the two-wheeled luggage barrows was the iron-shod front supports. These, set down on the cobbles, produced a terrific grinding noise, which caused a widespread panic among the timid and struggling passengers. The ascent of these barrows was aided on debarkation by great Clydesdale horses with clanking chains, which charged among the sea-sick and heavily-laden crowd. At the end of the summer season, when the trains were crowded by a southward-tending flight of passengers, the railway companies were frequently obliged to leave the baggage behind, and travellers to England found on arrival at Edinburgh that they must either stay the night or go on without their luggage. The distance across was about five miles, and on some days of wild weather the ferry service had to be suspended. The steamers were not large, as the harbour accommodation was not extensive, and in an easterly gale a crowded passage in the *John Stirling* or the *William Mure* was a most painful experience. In the dim distance, some eight miles away, the passenger could see and follow with his prayers the building of the great bridge which was to take him in safety and comfort from shore to shore.

The bridge, in fact, was urgently desired by the great railway companies, the local traders, and by the growing flight of passengers to Fife and the north. With the successful erection of the second Tay Bridge, which preceded the Forth Bridge by some years, the important trading centres of Dundee, Arbroath, and Aberdeen stood ready to add their traffic to the new route.

The elaborate history of the Forth Bridge,* re-

---

* We desire to acknowledge here our extreme indebtedness to this elaborate and beautiful volume. The following extract from a letter written by one of the editors, Mr. James Dredge, to Sir John Fowler, is of considerable interest :—

"*March 7th*, 1890. . . . I was pleased to hear from Percival this morning that you like our Record of the Bridge. I am well satisfied with it, and we all are. The opening of the Britannia Bridge, forty years ago, was celebrated by the production of a volume that cost £12,000 to produce ; the record of your work (so far) is a sixpenny newspaper !"

printed from *Engineering*, to which we are indebted for many of these particulars, tells of one or two earlier but abortive schemes for bridging the firth. A double tunnel was proposed in 1805. In 1818 Mr. James Anderson proposed a chain bridge, of which Mr. Westhofen has remarked that it was so slight a structure that it would hardly have been visible on a dull day, and that after a storm it would never again have been visible even on a clear day.

In 1860 and in 1865 Mr. (afterwards Sir Thomas) Bouch, at the instance of the North British Railway Company, proposed a bridge to cross at a point some six miles above Queensferry. This proposal fell through, and it was not till 1873 that a separate and independent company, the Forth Bridge Company, was formed to carry out Sir Thomas Bouch's design of a suspension bridge, with two spans of 1,600 feet each. The Great Northern, the North Eastern, the Midland, and the North British Railways agreed to find the capital, and to send traffic sufficient over the bridge to pay 6 per cent. on the contract price. Some foundation stones were laid, but in December, 1879, the first Tay Bridge, the design of Sir Thomas Bouch, collapsed, and the public confidence was shaken. Sir John Fowler, it may be remarked in passing, had always distrusted this work, and would not allow his family to cross it. One of the defects of the bridge was that the piers were constructed with too narrow a base. Sir John Fowler, on the day of the news reaching London, happened to meet his friend, Mr. Nasmyth, the inventor of the steam hammer, at an exhibition of Holbein pictures, and the remark passed between the two engineers that the ill-fated bridge might still be standing if the designer had adopted the Holbein "straddle," that peculiar attitude which the artist gives to his male figures, very particularly, it may be remembered, in his pictures of King Henry VIII. The design of the bridge, however, is now admitted to have been faulty in other respects. As a result of this mishap, Sir Thomas Bouch's design for a suspension bridge over the Forth was abandoned, and Messrs. Barlow, Harrison, and Fowler, the consulting engineers of the promoting railway companies, met to decide what steps should be taken. The inquiry considered the situation from every point of view. They discarded the proposal for a tunnel and the principle of a suspension bridge, and finally adopted, with some modifications introduced to meet the views of the other engineers, a plan submitted by Messrs. Fowler and Baker, who were appointed engineers to carry it into effect.

In July, 1882, an Act was obtained. Each of the contracting railway companies agreed to find its share of capital, and to guarantee the payment of 4 per cent. The North British Railway undertook to keep up the permanent way and to manage the traffic; while the Forth Bridge Railway Company was charged with the maintenance and repair of the fabric of the bridge.

With regard to the modifications introduced into the original design, Sir Benjamin (then Mr.) Baker, speaking at the Society of Arts shortly before the opening of the bridge, remarked :—

"If we had to do the work again, with our present experience, I doubt if some of these modifications would be insisted upon, whilst others would be made."

The spot chosen for the bridge is at Queensferry, where half-way across the channel of the firth stands the rocky island of Inchgarvie. On each side of the island the water is in places over 200 feet deep, and it was judged necessary to avoid this deep water and to build the bridge of two spans, with a central pier on the rock of Inchgarvie. Each span is of the unprecedented size of 1,710 feet, that is, three and two-third times as long as the Britannia Bridge over the Menai Straits, till then the longest span in the kingdom. The dimensions of each span is such that if erected in the Strand, near Charing Cross, its other end would reach across the Adelphi, the Thames Embankment, and the river itself, to the Surrey side.

The principle adopted is now known as that of the cantilever. With regard to this, Sir B. Baker has remarked :—

"When I was a student, a girder bridge which had the top member in tension and the bottom member in compression over the piers, was called a 'continuous girder bridge.' The Forth Bridge is of that type, and I used to call it a continuous girder bridge; but the Americans persisted in calling all the bridges they were building on the same plan 'cantilever bridges.' . . . Cantilever is a two-hundred-year-old term for a bracket, and the Forth Bridge spans are made up of two brackets and a connecting girder. Imagine two men trying to shake hands across a stream a little too wide for their hands to meet. One man extends his walking-stick, and the other grasps it, and so the stream is bridged. There we have the two arms or brackets and the connecting girder. In the Forth Bridge the arms are supported by great struts, as in a living model (shown in the illustration), where raking struts extended from the men's wrists to the points of support. The principle of bracket and girder construction is as old as the hills, for it lends itself particularly to timber construction, which we know in primitive times preceded masonry."

The living model mentioned above was arranged as follows :—

"Two men sitting on chairs extended their arms and supported the same by grasping sticks abutting against the chairs. This represented the two double cantilevers. The central beam was represented by a short stick slung from the near hands of the two men, and the anchorages to the cantilevers by ropes extending from the other hands of the men to a couple of piles of bricks. When stresses are brought to bear on this system by a load on the central beam, the men's arms and the anchorage ropes came into tension and the sticks and the chair-legs into compression." *

In the Forth Bridge it is to be imagined that the chairs are placed a third of a mile apart; that the men's heads are 340 feet above the ground; that the pull on each arm is about 4,000 tons, the thrust on each stick over 6,000 tons, and the weight on the legs of the chair about 25,000 tons.

The point most important to be grasped by the amateur is that each bracket or limb of the cantilever is an independent or self-supporting structure, and that in the connecting girder there is nothing of the nature of the keystone or locking of an arch. A neglect of this point, which is obvious enough, has given rise to some very inept criticisms.

In connection with the antiquity of the principle, Lord Napier of Magdala remarked to one of the engineers, "I suppose you touch your hats to the

---

* This illustration, originally devised by Sir B. Baker for the purpose of a popular lecture, has been reproduced in descriptions of the Forth Bridge in every language of the civilised globe. It may add to the cosmopolitan interest of the illustration if it is stated that the central figure is Mr. Kaichi Watanabé, then an engineering student with Messrs. Fowler and Baker, and now President and Engineer-in-Chief of several Japanese railways.

A LIVING MODEL OF THE FORTH BRIDGE.

Illustrating the principle of the cantilever.

To face page 282

Chinese?" "Yes, indeed," was the reply, "bridges on that principle were built in China many centuries ago." The bridges familiar to us in the old willow-pattern crockery are for the most part of this design. An even older anticipation of the tubular-bridge principle (which consists, to speak popularly, in dividing the girder into an upper and lower member, by taking away the material at the centre where it is not required, and adding it where it is most wanted to resist the strain of bending, *i.e.* by tension in the upper and compression in the lower member of the so-called tube or girder) has been noted by Dr. McAlister in a most interesting article entitled "How a Bone is Built," published in the *English Illustrated Magazine* for July, 1885, in which he compares the deft and economical process of nature in the fashioning of a bone with the practice of the engineers who built the Britannia Tubular Bridge, and the great Forth Bridge then in process of construction.

The merit of the Forth Bridge lies not in its originality, but in the successful application of an old principle to new conditions. In an article contributed to the *Nineteenth Century* review, July, 1889, Sir John Fowler and Sir Benjamin Baker wrote :—

"The adaptability of the cantilever system of construction for railway bridges of large span became obvious to ourselves, and no doubt to others, soon after the invention of Bessemer made cheap steel a possibility. In 1865 we designed a steel cantilever bridge of 1,000 feet span for a proposed viaduct across the Severn, near the site of the present tunnel ;* but it was not until 1881 that the Forth Bridge designs were published in the English and American technical journals.

* With regard to this proposal, see p. 356.

These designs naturally attracted much attention, and with characteristic promptness American engineers realised the advantages of the system, and designed and built the following year a steel cantilever railway bridge on the Canadian Pacific Railway, and have since followed on with more than half a dozen others of the same type of construction."

The history of this great engineering feat is in itself a beautiful illustration of the relation of human effort or ability to the environment by which it is necessarily surrounded. The "able men" who designed and carried into effect this stupendous structure have been themselves most insistent in giving prominence to the contributing influence of outside conditions. The antiquity of the principle, and the new value given to it by the invention of new processes for the cheap production of suitable material, have been amply noticed in the above quotation, but the most important consideration of all is that set out by Sir Benjamin Baker in his presidential address to the Institution of Civil Engineers in 1895. He there draws attention to what we may term the economic as opposed to the material conditions which are now necessary to the successful accomplishment of a great design.

"An impartial survey of actions and events recorded in history will satisfy most people that in all ages there were to be found men no less intellectual and enterprising than ourselves, and that the demands of the time, whether warlike or peaceful, have always proved capable of realisation. That 'Necessity is the mother of Invention' is true of all ages, and the only difficulty in making forecasts is the strange way in which the aims of different generations vary. . . . The popular notion that some great advance is due to the brilliant inspiration of a particular genius proves, on closer examination, to be wrong, as the advance was merely the result of the operation of the ordinary laws of supply and demand, and the genius himself very probably will have committed himself in writing to a sufficient extent to prove that he really was drifting with the stream rather than piloting the ship."

Without accepting in full this modest disclaimer of originating ability, we are enabled, by the consideration here put forward, to estimate in their true proportions the various currents which unite to form the full stream of progress. Originating ability is controlled and directed by the demand of the time; till demand arises, it is apt to lie dormant or to expend itself on other objects. Thus the workmen who built the *Santa Maria* for Columbus, and the *Royal Harry* for our Tudor king,

" were quite capable as artificers of constructing with the same materials and implements clipper ships of from 500 to 900 tons, such as astonished the world by the historical race from China to London in 1866."

The reason, that ingenuity in those earlier times lay comparatively idle, is to be found in the fact that

" wars, revolutions, and great social changes occupied men's thoughts in former times, and there was not that unceasing struggle for commercial supremacy and material advantages which is so characteristic of the present century."

Apart, therefore, from the ability of the builders, and the presence of suitable material at a moderate cost, the Forth Bridge is also due to those industrial conditions which make the saving of a few hours' time on the transport of passengers and goods a motive of irresistible urgency.

A bridge on this stupendous scale, if it was to secure the public confidence, must possess great rigidity. It must not only be able to carry the burdens imposed upon it, but, if the timidity of the travelling public was to be overcome, vibration both under the load of the passing trains and also from the lateral pressure of the wind must be reduced to an imperceptible minimum. Further, if it was to be built at all, the incomplete structure must (as the period of erection was to last some years) be as competent to resist the force of the hurricane as the finished bridge. Nothing but the best material could be used, yet at the same time a reasonable economy (as far as was compatible with these requirements) had to be observed, in order to bring it within the category of a legitimate commercial venture.

In securing these imperative conditions the distribution of the weight of the bridge is of the first importance. The design adopted throws the greater proportion of the weight, nearly one-fourth of it, immediately over the main piers or supports, where it is, of course, most easily provided for. It also is so devised that at the points where the lateral pressure of the wind is most to be dreaded, the smallest surface is presented to the blast. Thus at the central tower of the Inchgarvie pier the weight per foot run is 23 tons, and, in the first bay of the cantilever, 21 tons; while on the central girders of 350 feet span, which join the arms of the cantilevers, that is at the point furthest from the main support, the weight is only a little over 2 tons per foot. Again, the surface presented to the wind is rapidly lessened as the brackets recede from the massive piers, and on the centre of the span where the leverage of wind pressure is most formidable, the narrowest surface is exposed. Great stability is also given to the structure by the "straddling" of the supporting columns.

Again, during the course of construction the several portions of the bridge were placed securely in position, one after the other; there was little or no temporary work required, and each addition to the structure served as a staging for the work that was to follow.

A further difficulty was occasioned by the fact that the enormous mass of metal employed in the building is liable to contraction and expansion from changes of temperature. This has been adequately provided for by expansion-joints, a device readily adaptable to the cantilever and central girder principle of construction.

On December 21st, 1882, the contract for the construction of the Forth Bridge was let to the firm of Messrs. Tancred, Arrol, and Co. The original contract price of £1,600,000 and the specified time were inevitably exceeded. The total expenditure on the Forth Bridge itself, and on the new railways connecting it with the North British Railway system, was £3,367,625, and the net earnings suffice to pay four per cent. upon this amount.

Let us begin our description of the famous structure by giving an account of the supporting piers.

The point chosen for the bridge is at a narrowing of the firth caused by a promontory on the north or Fife side. This projection of land reduces the crossing of the river to 1 mile and 150 yards. The promontory is of hard whinstone. Midway in the channel, at a distance of one-third of a mile from the top of the promontory, and due south, lies the rock of Inchgarvie. The north channel, between the island and the Fife coast, has a depth of about 200 feet, and is the one generally followed by the shipping.

The southern edge of the island is under water, and from it to South Queensferry is a distance of 2,000 feet, of which about 500 feet are uncovered at low tide. The whinstone, or basaltic trap rock, which forms the foundation of the great Fife and of the Inchgarvie piers, disappears in the southern channel (*i.e.* between South Queensferry and Inchgarvie), and is overlaid by a thick bed of hard boulder clay with a forty-feet topping of softer clay, silt, and gravel. When the southern mainland is reached, the ground rises gradually, the clay disappears, and ledges of freestone rock crop out. The foundations of the third or South Queensferry Pier are laid in this bed of boulder clay. From an engineering point of view the nature of the foundations has been considered highly satisfactory, and no trouble was at any time experienced in this respect.

The following is taken from Sir B. Baker's speech at the Royal Institution on May 20th, 1887 :—

"The total length of the viaduct is about 1½ miles, and this includes two spans of 1,700 feet, two of 675 feet being the shoreward ends of the cantilever, and fifteen of 168 feet. Including piers there is thus almost exactly one mile covered by the great cantilever spans, and another half-mile of viaduct-approach. . . .

"Each of the main piers includes four columns of masonry founded on the rock or boulder clay. Above low water the cylindrical piers are of the strongest flat bedded Arbroath stone set in cement and faced with Aberdeen granite. The height of these pieces of masonry is 36 feet, and the diameter 53 feet at bottom and 49 feet at top, and they each contain

48 steel bolts 2½ inches in diameter and 24 feet long to hold down the superstructure.

"Below low water the piers differ somewhat in character, according to the local conditions. On the Fife side, one of the piers was built with the aid of a half-tide dam, and the other with a full-tide dam. The rock was blasted into steps, diamond drills and other rock drills being used. Even this comparatively simple work was not executed without considerable trouble, as the sloping rock bottom was covered with a closely compacted mass of boulders and rubbish, through which the water flowed into the dam in almost unmanageable quantity. After many months' work the water was sufficiently excluded, by the use of cement bags and liquid grout poured in by divers under water and other expedients, and the concrete foundation and masonry were proceeded with."

For the sake of those uninitiated in the use of engineering terms, a word may here be interpolated as to "dams," or "cofferdams," as they are sometimes called, and as to "pneumatic caissons,"—two devices for facilitating foundation-work in water. These, of necessity, played a most important part in the construction of this bridge. The following is abridged from the explanation contained in the reprint from *Engineering*. A cofferdam or caisson may be described as an enclosure in water for the purpose of laying dry the space enclosed, or at any rate of preventing a flow of water through it. In soft ground this is done by driving a double row of piles at a distance of from two feet to four feet from each other, till a double timber wall exists all round. Sluice doors or valves are placed so as to allow the tide to flow in and out. The single timber piles are held together by longitudinal timbers aided by stays and struts to resist the water pressure without. The space between the two lines

U

of piles is then cleared out, and filled in with clay puddle, and pressed down, till the whole is filled in to full-tide or half-tide level, as the case may be. The sluices are then closed and the water pumped out, the bottom is thus exposed for examination or for foundation work. The choice between a full-tide or half-tide dam will be generally determined by the firmness of the hold obtained by the piles, and also by the extent of external pressure from water, tide, and wind. Into a half-tide dam the water must be allowed to enter, when the tide rises to the level of the dam wall, and requires to be pumped out when the tide again falls. With a full-tide dam the water is permanently excluded and work can go on continuously.

When working on rock the driving of piles is impracticable, and other means have to be devised. Dams giving access to the proposed foundations had in such cases to be made by sinking shields fitted to the contour of the rock, aided by submarine building operations by divers, who laid bags of concrete in position, and so gradually built up half-tide caissons such as were used at the Inchgarvie north circular piers. As an example of work of this character it may be stated that at the north-east pier there were two half-tide caissons, and out of them there had each day to be pumped 250,000 gallons in one case and 340,000 in the other. The time occupied was just under an hour.

Finally, if instead of a caisson open at the top the caisson is covered in, like a bell or gas-holder, and the water is driven out by forcing air in, thereby allowing the workmen to enter and excavate in the dry, it is a caisson worked by the pneumatic process.

When the foundations, as at the southernmost piers of Inchgarvie, had to be constructed in deep water this last method became necessary.

"Several designs," says Sir B. Baker in his Royal Institution lecture, "were prepared for these foundations, but it was finally decided, and as experience proved wisely, to put them in by what is known as the pneumatic or compressed-air process. The conditions of the problem were a sloping, very irregular, and fissured rock bottom, in an exposed sea-way, and with a depth at high water of 72 feet. Anything in the nature of a water-tight cofferdam, such as used at the shallow piers, was out of question, and the plan adopted was as follows:—

"Two wrought iron caissons, which might be likened to large tubs or buckets, 70 feet in diameter and 50 to 60 feet high, were built on launching-ways on the sloping southern foreshore of the Forth. The bottom of each caisson was set up 7 feet above the cutting edge, and so constituted a chamber 70 feet in diameter and 7 feet high, capable of being filled at the proper time with compressed air, to enable the men to work, as in a diving-bell, below the water of the Forth. The caisson weighed about 470 tons, was launched, and then taken to a berth alongside the Queensferry jetty, where a certain amount of concrete, brickwork, and staging was added, bringing the weight up to 2,640 tons. At Inchgarvie a very strong and costly iron staging had previously been erected, alongside which the caisson was finally moored in correct position for sinking. Whilst the work described was proceeding, divers and labourers were engaged in making a level bed for the caisson to sit on. The 16-feet slope in the rock bottom was levelled up by bags filled with sand or concrete. As soon as the weight of the caisson and filling reached 3,270 tons the caisson rested on the sand bags and floated no more. The high ledge of rock upon which the northern edge of the caisson rested was blasted away, holes being driven by rock drills and otherwise under the cutting edge and about six inches beyond for the charges. After the men had gained a little experience in this work no difficulty was found in undercutting the hard whinstone rock to allow the edge of the caisson to sink, and of course there was still less difficulty in removing the sand bags temporarily used to form a level bed. The interior rock was excavated as easily as on dry land, the whole of the 70 feet diameter by 7 feet high chamber being thoroughly lighted by electricity. Access was obtained through a vertical tube with an air-lock at the top, and many visitors ventured to pass through this lock into the lighted chamber below, where the pressure at times was as high as 35 lbs. per square inch. Probably the most astonished visitors were some salmon, who, attracted by the commotion in the water caused by the escape of compressed air under the edge of the caisson, found themselves in the electric-lighted chamber. When in the chamber the only notice of this escape of large volumes of air was the sudden pervadence of a dense fog, but outside a huge wave of aerated water would rise above the level of the sea, and a general effect prevail of something terrible going on below. No doubt the salmon thought they had come to a cascade turned upside down, and, following their instinct of heading up it, met their fate. Another astonished visitor was a gentleman who took a flat-sided spirit flask with him into the caisson, and emptied it when down below. Of course the bottle was filled with compressed air, which exploded when passing through the air-lock into the normal atmospheric pressure, the pressure in the bottle being 33 lbs. per square inch.

"The Garvie piers, notwithstanding the novelties involved in sinking through whinstone rock at a depth of 72 feet below the waves of the Forth, were completed without misadventure in less than the contract time. The first of the deep Garvie caissons was launched on March 30th, 1885, and both piers were finished to sea-level or above by the end of the year.

"At Queensferry all four piers were founded on caissons identical in principle with those used for the deep Garvie piers. The deepest was 89 feet below high water, and weighed 20,000 tons; the shallowest of the four was 71 feet high, the diameter in all cases, as at Garvie, being 70 at the base. Some difference in detail occurred in these caissons, as compared with Garvie, owing to the differences of the conditions. Thus, instead of a sloping surface of rock the bed of the Forth was of soft mud to a considerable depth, through which the caissons had to be sunk into the hard boulder clay. Double skins were provided for the caissons, between which concrete could be filled in to varying heights if necessary, so that greater weight might be applied to the cutting edge where the mud was hard than where it was soft. The annular wall of concrete also gave great strength to resist the hydrostatic pressure outside the caisson, for it must be understood that the water was excluded both below and above the working chamber.

"The process of sinking was as follows: the caisson being seated on the soft mud, which, of course, practically filled the working chamber, air was blown in, and a few men descended the shaft or tube of access to the working chamber in order to clear away the mud. This was done by diluting it to the necessary extent by water brought down a pipe under pressure, and by blowing it out in this liquid state through another pipe by means of the pressure of air in the chamber. It was found that the mud sealed the caisson, so that a pressure of air considerably in excess of that of the water outside could be kept up, and it was unnecessary to vary the pressure according to the height of the tide. In working through the soft mud both intelligence and courage were called for on the part of the men, and it is a pleasure and duty for me to say that the Italians and Belgians engaged on the work were never found wanting in those qualifications. . . .

" With one of our caissons we unfortunately had an accident and loss of life. On New Year's Day, 1885, the south-west Queensferry caisson, which had been towed into position and weighted with about 4,000 tons of concrete, stuck in the mud, and instead of rising with the tide remained fixed, so that the water flowing over the edge filled the interior. The additional weight of water caused the caisson to sink further in the mud, especially at the outer edge, and to slide forward and tilt. The contractors determined to raise the skin of the caisson until it came above water-level, and then pump out and float the caisson back into position. About three months were occupied in doing this, but when pumping had proceeded a certain extent the caisson collapsed, owing to the heavy external pressure of the water, and two men were killed. It was necessary then to consider very carefully what had better be done, as the torn caisson was difficult to deal with."

Some arduous work had to be done by the divers to make the caisson again watertight.

"Finally, on October 19th, 1885, or between nine and ten months after the first accident, the caisson, to the relief of everyone, was floated into position, and the sinking proceeded without further difficulty. Thus the last of the main piers was completed in March, 1886, almost exactly two years after the first caisson was floated out."

It was necessary to restrict the hours of work in these compressed-air chambers, as the men suffered more or less from pains in the limbs and elsewhere. Paralysis and other serious consequences are reported to have followed on too long an exposure to these conditions. There were, however, at the Forth Bridge no deaths directly resulting from the air pressure, and Sir Benjamin Baker says that he himself suffered no sort of inconvenience from entering the chamber.

For a description of the superstructure on the main piers, which we must now suppose to be successfully founded, we must rely again on Sir B. Baker's lecture at the Royal Institution.

Owing to the unprecedented length of the span, the dead weight of the structure itself was far in excess of any number of railway trains which could be brought upon it. Thus the weight of one of the 1,700-feet spans is about 16,000 tons, while the heaviest rolling load could not be more than 800 tons,* or only 5 per cent. of the dead weight.

"It is hardly necessary to say that the bridge will be as stiff as a rock under the passage of a train. Wind, however, is a more important element than train weight, as with the assumed pressure of 56 lbs. per square foot the estimated lateral pressure on each 1,700-feet span is 2,000 tons, or two and a half times as much as the rolling load. To resist wind the structure is 'straddle-legged,' that is, the lofty columns over the piers are 120 feet apart at the base and 33 feet at the top. Similarly the cantilever bottom members widen out at the piers. All of the main compression members are tubes, because that is the form which, with the least weight, gives the greatest strength. The tube of the cantilever is at the piers 12 feet in diameter and $1\frac{1}{4}$ inch thick, and it is subject to an end pressure of 2,282 tons from the dead weight, 1,022 tons from the trains, and 2,920 tons from the wind, total 6,224 tons, which is the weight of one of the largest trans-atlantic steamers with all her cargo on board. The vertical tube is 343 feet high, 12 feet in diameter, and about $\frac{5}{8}$ inch thick, and is liable to a load of 3,279 tons. The tension members are of lattice construction, and the heaviest stressed one is subject to a pull of 3,794 tons. All of the structure is thoroughly braced together by wind bracing of lattice girders, so that a hurricane or cyclone storm may blow in any direction up or down the Forth without affecting the stability of the bridge. Indeed, even if a hurricane were blowing up one side of the Forth and down the other, tending to rotate the cantilevers on the piers, the bridge has the strength to resist such a combination."

The highest recorded pressure on the wind gauges on Inchgarvie during the period of construction would

---

* The weight actually put on it by the Government inspectors in their last test consisted of two trains, each weighing about 901 tons.

not, in Sir B. Baker's opinion, have averaged more than 20 lbs. per square foot. With regard to this question of wind pressure, it is interesting to notice that Sir Thomas Bouch, when asked at the Court of Inquiry why he had made the unfortunate Tay Bridge so much weaker than other tall viaducts erected by himself, replied that his ideas on the subject of wind pressure had been modified by a statement of Sir George Airy, the late Astronomer Royal, contained in a report on the proposed Forth Bridge to the effect that "the greatest wind pressure to which a plane surface like that of the bridge will be subjected in its whole extent is 10 lbs. per square foot." The Board of Trade, however, warned by the fate of the Tay Bridge, had laid it down that a 56-lbs. pressure must be provided for, while the technical work of Tredgold named 40 lbs. as the limit. Sir George Airy, somewhat incautiously, published in the columns of *Nature*, October 19th, 1882, an adverse criticism of the Fowler-Baker design, and endeavoured to show that a suspension bridge would have been more suitable. Sir George Airy was then over eighty years of age, but his high repute as a man of science gave weight to his opinion, and his criticisms caused uneasiness to the uninstructed public and considerable indignation in the technical journals. One of them, *Engineering*, December 15th, 1882, rather cruelly set out a long list of similar ineptitudes committed by this distinguished astronomer, whose adventures in engineering criticism provide an almost heroic instance of the soundness of the maxim, *ne sutor ultra crepidam*.

"Sir George Airy's periodic attacks on engineering works are instructive, inasmuch as they serve to indicate the blunders which inexperienced engineers would be likely to fall into, if entrusted with responsibility before being properly qualified. The late Astronomer Royal throughout his long career has laboured under the hallucination that the science of engineering is not a matter of experience and research, but of intuition. . . . Possibly he may have at times mistaken the silence of engineers for acquiescence, but supposing Sir George wrote a letter, stating that in his opinion the distance from London to Edinburgh was 20 miles, we doubt whether anyone would take the trouble to correct him."

On this occasion, however, several distinguished engineers took part in the controversy, and showed very conclusively the error and irrelevancy of Sir G. Airy's criticisms.

To return to the question of the manufacture of the plates which composed the tubes used in the bridge. Special plant had to be devised for their preparation. Long furnaces, heated in some instances by gas producers and in others by coal, first heated the plates, which were then hauled between the dies of an 800-ton hydraulic press, and bent to the proper radius. When cool, the edges were planed all round, and the plates built up into the form of a tube in the drilling yard. Here they were dealt with by eight great travelling machines, having ten traversing drills radiating to the centre of the tube, and drilling through as much as four inches of solid steel in places. When complete, the tubes were taken down, the plates cleaned and oiled and stacked ready for erection. The magnitude of the work justified the erection of the most elaborate and complete workshops on the Queensferry shore. Here the whole of the steel work of the bridge was made and fitted. This fact should be noted as one of the many unique features in the building of the Forth Bridge.

"The tension members and lattice girders"—we quote again Sir B. Baker—"generally are of angle bars, sawn to length when cold, and of plates planed all round. Multiple drills tear through immense thickness of steel at an astonishing rate. The larger machines have ten drills, which, going as they do, day and night, at 180 revolutions per minute, perform work equivalent to boring an inch hole through 280 feet thickness of solid steel every twenty-four hours. About four per cent. of the whole weight of steel delivered at the works leaves it again in the form of shavings from planing machines and drills. The material used throughout is Siemens steel of the finest quality, made at the Steel Company's works in Glasgow, and at Landore in South Wales. Although one and a half times stronger than wrought iron, it is not in any sense of the word brittle, as steel is often popularly supposed to be, but it is tough and ductile as copper. You can fold half-inch plates like newspapers, and tie rivet bars like twine into knots."

As already explained, owing to the depth of water and to the exposed situation in the open firth, scaffolding was impossible; the bridge had, therefore, to constitute its own scaffolding.

"The principle of erection adopted was, therefore, to build first the portion of the superstructure over the main piers, the great steel towers, as they may be called, although really parts of the cantilever, and to add successive bays of the cantilever right and left of these towers, and therefore balancing each other, until the whole is complete. This being the general principle, a great deal yet remained to be done in settling the details."

What was finally settled has been described as follows :—

"After the skewbacks, horizontal tubes, and a certain length of the verticals, as high as steam cranes could conveniently reach, were built, a lifting stage was erected. This consisted of two platforms, one on each side of the bridge, and four hydraulic lifting rams, one in each 12-feet tube. To carry these rams, cross girders were fitted in the tubes, capable of being raised so as to support the rams and platforms as erection proceeded, and steel pins were slipped in to hold the cross girders. Travelling cranes are placed on the platforms, and these cranes, with the men working aloft, are, of course, raised with the platforms, when hydraulic pressure is let into the rams. The mode of procedure is to raise the platform one foot, and slip in the steel pins to carry the load, whilst the rams are getting ready to make another stroke of one foot. When a 16-feet lift has been so made, which is a matter of a few hours, a pause of some two or three days occurs to allow the riveting to be completed. The advance at times has been at the rate of three lifts, or 48 feet in height in a week.

"The riveting appliances designed by Mr. Arrol are of a very special and even formidable character, each machine weighing about 16 tons. It consists essentially of an inside and outside hydraulic ram mounted on longitudinal and annular girders in such a manner as to command every rivet in the tubes and to close the same by hydraulic pressure. Pipes from the hydraulic pumps are carried up inside the tubes to the riveters, and oil furnaces for heating the rivets are placed in convenient spots also inside the tubes. By practice and the stimulus of premiums the men have succeeded in putting in 800 rivets per day with one of the machines at a height of 300 feet above the sea, which, in fact, is more than they accomplished when working at ground level. Indeed, by the system of erection adopted, the element of height is practically annihilated, and with ordinary caution the men are safer aloft than below, as in the former case they are not liable to have things dropped on their heads."

In this way the central steel towers were built. The successive bays of the cantilever had then to be added.

The first half-bay of each cantilever was erected by a similar use of platforms and overhead cranes.

"Our experience so far," said Sir B. Baker in September, 1889, "has been that whilst the work of erection has been somewhat slower and more costly than we anticipated, it has, on the other hand, been less difficult and subject to fewer contingencies. We thought at first that the crane men and erectors would require practically to be close together, but we have found out, or rather the men have found out for themselves, that cranes 370 feet up in the air can handle work at ground level, and that the long steel wire ropes hanging from the crane jibs, instead of being destructive of their usefulness, are often of great advantage, as plates and bars can be swung out pendulum fashion to a distance far beyond the reach of the jib itself. This result of experience, combined with the boldness of the men, enables us to dispense with the use of lifting platforms for the outer bays of the cantilevers, and in lieu of this mode of erection to use steam cranes travelling on the top. . . ."

The girders which were to carry the permanent way and the trains were put in when the pile rose to the necessary height, and furnished an additional platform for work on the still uncompleted parts of the structure.

"The viaduct girders," says *Engineering*, "were now built out by overhanging into the cantilevers. . . . The viaduct girders were strong enough to carry themselves overhanging for a distance of 100 feet, and even then to carry at the forward end the weight of a 3-ton crane and its load; but, as matter of safety, wire ropes were carried from the outer ends up to the top junctions on the vertical columns."

INCH GARVIE MAIN PIER.
South Cantilever, Sept. 1888.

To face page 300.

We have now to describe the joining of the cantilevers by the central girders, the crowning feat, and completion of the whole structure. Owing to the swiftness of the current and the exposed nature of the work, it was not possible to hoist the girder into position, and it was decided to build it out piece by piece till the arms of the cantilevers met in the middle. It must be borne in mind, however, that, strictly speaking, the great 1,710-feet spans are not joined in the ordinary sense of the term. Each half-span hangs entirely from its own supports on the main piers. Owing to the large amount of expansion and contraction to which the immense mass of metal is liable, it would endanger the fabric were it actually joined. At the junction of the central girder with the cantilevers expansion-joints are introduced, so that the shrinkage due to the cold may not cause a gap, and so that the expansion due to the heat may not cause " buckling." The extreme variation in the lengths of the 1,710-feet spans under alternations of heat and cold is calculated not to exceed nine inches, but provision has been made for a variation of double that amount.

" Almost every engineering visitor to the works," says Sir B. Baker, in September, 1889, "during past years has asked, ' How are you going to erect the central girder ? ' I have never varied myself in opinion as to what would be the best way of doing the work. In a paper read before the British Association at Southampton, in 1882, or seven years ago, I said, ' The central girder will be erected on the overhanging system, temporary connection being formed between the ends of the cantilevers and central girders. The closing lengths or key-pieces at the centre of each 1,700-feet span will be put in on a cloudy day, or at night, when there is little variation of temperature, and the details will be so arranged that the key-piece can be completed and the temporary connections cut away in a few hours, so as to avoid any temporary inconvenience from changes of temperature.' That is a sufficiently concise description of the plan now in progress for erecting the central girders."

During the construction of the works Mr. Arrol showed an interesting model, illustrating how the expansion and contraction of the metal mass was provided for at the junction of the cantilevers and the girders. A movement of six inches was provided for at the end of each cantilever. The contrivance to admit of it is most clever. What is termed a rocking-pillar is introduced between the end of the cantilever and the end of the girder. The lower end of the pillar rests in a socket on the cantilever, while the upper end supports a socket carrying the girder. In this way a certain amount of play is given to the junction. The rocking motion may be illustrated by placing a walking-stick on the ground, and allowing the upper end to oscillate an inch or so. In this case the ground would represent the cantilever, and the hand at the top the girder attachment. Other arrangements providing against undue strain owing to contraction and expansion were introduced in the shape of "sliding bed-plates" and "roller-bearings."

The following account of the actual joining of a girder on November 14th, 1889, must conclude our description of this great work. It is taken from the already frequently quoted pages of *Engineering*.

"The north central girder had in the meantime been built out in a precisely similar manner, and by October 15th it was sufficiently advanced to allow a gangway, 65 feet long, to be laid across. This enabled the directors of the company

NEARING COMPLETION.

From the South-east, September 17, 1889.

THE FORTH BRIDGE.

From the South-east, November 1, 1889.

To face page 303.

to walk across the bridge from end to end, the chairman of the company being actually the first person to cross the north span. By October 28th the last booms were put in, and by November 6th everything was ready to connect the girder also. The temperature on that day did not rise, however, sufficiently high to make the joint, but in the night a sudden rise took place, and by 7.30 in the morning the bottom booms were joined together for good.

"It now required a good fall of the temperature to get the top booms connected, for the two halves of this girder had been set less high at starting, and there was now practically no camber in the bottom booms. But the weather remained obstinate and the temperature very high, and it was not until the morning of November 14th that the key-plates could be driven in, and the final connection made. An episode, of which much has been made in the papers, occurred on this occasion, and the facts are simply as follows. After the wedges at the bottom ends had been drawn out and the key-plates driven in, a slight rise of temperature was indicated by the thermometer in the course of the morning, and orders were given to remove the bolts in the central joints of the connecting ties and to light the furnaces. Whether the thermometer indicated wrongly, or whether the cantilevers had not had time to fully expand under the rise of temperature, or whether a decrease of the same took place, it is not now possible to prove, but when only about thirty-six of the turned steel bolts remained in the joints, and before the furnaces could get fairly started, the plate-ties sheared the remaining bolts and parted with a bang like a shot from a 38-ton gun. Something of a shake occurred in the cantilevers, which was felt at the opposite ends and caused some little commotion among the men. No mishap occurred, however, and nothing in the way of a fall of the girders took place as stated in the papers, simply the work of the furnaces and the task of knocking out thirty-six bolts was saved, and the girder swung in its rockers as freely as if it had been freed in the most natural manner.

"And thus the Forth Bridge was completed, for the remaining work was simply to replace temporary connections by permanent ones, to rivet up those which were only bolted, and do the thousand and one things which always remain to be done after everything is said to be finished."

With regard to the staff of workmen, we may again quote Sir B. Baker :—

"To carry out the work at the Forth Bridge there is an army of 3,500 workmen, officered by a proportionate number of engineers. Everything, except the rolling of the steel plates, is done on the spot, and consequently there are literally hundreds of steam and hydraulic engines and other machines and appliances too numerous to mention, many of them of an entirely original character.

"It is, of course, impossible to carry out a gigantic work of this kind . . . without paying for it, not merely in money but in men's lives. I shall have failed in my task if you do not to some extent realise the risks to which zealous and plucky workmen will be sure to expose themselves in pushing on with the work of erecting the Forth Bridge. Speaking on behalf of the engineers, I may say that we never ask a workman to do a thing which we are not prepared to do ourselves, but of course men will on their own initiative occasionally do rash things. Thus, not long ago a man trusted himself at a great height to the simple grasp of a rope, and his hand getting numbed with cold he unconsciously relaxed his hold and fell backwards a descent of 120 feet, happily into the water, from which he was fished out, little the worse, after sinking twice. Another man, going up in a hoist the other day, having that familiarity with danger which breeds contempt, did not trouble to close the rail, and stumbling backwards fell a distance of 180 feet, carrying away a dozen rungs of a ladder with which he came in contact as if they had been straws. These are instances of rashness, but the best men run risks from their fellow-workmen. Thus, a splendid fellow, active as a cat, who would run hand over hand along a rope at any height, was knocked over by a man dropping a wedge on him from above, and killed by a fall of between one and two hundred feet. There are about 500 men at work at each main pier, and something is always dropping from aloft. I saw a hole one inch in diameter made through the four-inch timber of the staging by a spanner which fell about 300 feet, and took off a man's cap in its course. On another occasion a dropped spanner entered a man's waistcoat and came out at his ankle, tearing open the whole of his clothes but not injuring the man himself in any way."

To give some idea of the size of this work, it is calculated that there are about eight million of rivets to be driven. The amount of surface to be painted is equal to 145 acres. The weight of steel in the main spans is 51,000 tons. The weight of the 1,710 span is 16,000 tons. The steel plates required in the construction, if placed in line, would have stretched 45 miles. In the construction there was used 21,000 tons of cement, 47,000 tons of granite, and 113,000 tons of stone.

The directors of the Forth Bridge Railway Company were—Mr. M. W. Thompson (chairman) and Mr. W. Unwin Heygate, representing the Midland Railway; Lord Colville of Culross and Lord Hindlip from the Great Northern Railway; Mr. John Dent-Dent (deputy-chairman) and Sir Matthew White-Ridley, Bart., from the North Eastern Railway; the Marquis of Tweeddale and the Earl of Elgin and Kincardine from the North British Railway. Mr. Spencer Brunton and Mr. James Hall Renton were elected by the shareholders. The secretary was Mr. G. B. Wieland, the secretary of the North British Railway. The engineers were Sir John Fowler, K.C.M.G., C.E., and Mr. Benjamin Baker. The contractors for the bridge were Sir Thomas S. Tancred,

x

Bart., Mr. W. Arrol, Mr. T. H. Falkiner, and Mr. Joseph Phillips. The contractors for the north and south approach railways were Mr. W. Arrol, Mr. T. H. Falkiner, and Mr. Joseph Phillips. On the staff of Sir John Fowler and Mr. Baker were the following : Mr. Allan Stewart, Mr. P. W. Meik (resident engineer from, 1883 to 1886), Mr. F. E. Cooper (resident engineer from 1886 to 1890), and a number of assistants. On the contractors' staff were Mr. Thomas Scott, manager ; Mr. W. Westhofen, who was specially engaged on the works at Inchgarvie ; Mr. A. S. Biggart, in charge of drawing offices, shops, and yards ; and a number of others far too numerous to mention. M. Coiseau, a Belgian contractor, was in charge of the pneumatic caissons.

When the Forth Bridge was begun, Sir John Fowler was no longer a young man ; but though occasionally in his correspondence there is allusion to a troublesome bronchial affection, his energy and determination remained unabated. Nominally the chief responsibility rested on his shoulders, but of course the detailed plans of this gigantic undertaking required the co-operation of many minds. It is a structure which has made the reputation of many men. Beyond the general responsibility which fell on him as the titular chief of the engineering staff, Sir John's personal interest was principally attracted to the masonry of the piers and to the quality of the granite and other materials employed in them, a subject which always had had a great fascination for him, and when it is remembered that the approach viaducts on each side of the Forth are in themselves large engineering works involving a succession of great granite piers, the importance of Sir John's favourite study of masonry becomes apparent.

At the same time, his attention was not exclusively directed to this point.

The following passage from an address delivered by him to the Merchant Venturers' School at Bristol sums up in popular language the great change which, in the course of his own observation, had come over the business of bridge building, mainly by reason of improvements in material—a change which indeed seemed to culminate in the methods and designs used on the Forth Bridge :—

"If I were to be called upon to name one science which has contributed more than any other to the improvement of the quality of material, or to economy in its production, I think I should name chemistry. By means of chemistry efficiently applied, after thousands of experiments, we have steel of a high quality at a less cost per ton than iron of even moderate quality could formerly be produced.

"At the present time 14,000,000 tons of steel rails are annually made in the world at a less cost per ton, and of more than three times the durability (or length of life) of the rails formerly manufactured from the metal called 'iron.'

"What this means in annual economy and greater freedom from accidents would lead us, if we pursued the investigation, into very large figures and elaborate statistics, but it is obviously of such vast importance that it may be classed as one of the greatest scientific and practical improvements of modern times.

"Again, by the manufacture of steel plates for ships and bridges we have a vastly superior and stronger material at less cost than iron. . . . The Forth Bridge . . . is an excellent illustration of the value of the superior and economical steel material, first introduced into this country by the process called 'open hearth,' through the genius of my late distinguished friend Sir William Siemens. It is not too much to say that the great Forth Bridge would have been financially impossible without the use of steel. With iron it would have been twice the weight (if practicable at all), in consequence of the less strength of the material, and more than twice the cost, and therefore impracticable. . . .

" The Britannia Bridge over the Menai Straits was built forty years only before the Forth Bridge, but how great the contrast between the two works !

" Let us compare design, material, and manufacture.

" The design, in the hands of Stephenson and Fairbairn, was in accordance with the best knowledge and experience of that day, and presents a solid and safe structure, but with the material so placed that much of it contributes nothing to the strength of the bridge, and is mere dead weight, tending to weaken the bridge.

" On the other hand, the Forth Bridge is designed so that each member in the structure, either in tension or compression, has a definite and calculated stress to sustain.

" The material of which the Menai Bridge was constructed is iron, having an ultimate tensile strength of 22 tons per square inch, whilst the Forth Bridge is of steel, with a tensile strength of 33 tons per square inch.

" The improvement in manufacture has been as great as in the strength and excellence of the material. For instance, for the Menai Bridge, plates of the dimensions of 12 feet long by 2 feet wide, or 24 square feet in area, were obtained with difficulty, and special credit was accorded to Mr. Thorneycroft of Wolverhampton for his practical skill in obtaining plates of that large size. On the Forth Bridge the largest plate was 30 feet by 5 feet, or 150 square feet in area, and was obtained without difficulty or extra expense, and thus an immense number of rivets were saved.

" The engineer of the Menai Bridge calculated that, with the iron they used, a bridge of the span of 1,710 feet might be constructed, but not an ounce of weight must be put upon it, or a breath of air allowed to impinge against it. Each span of the Forth Bridge happens also be 1,710 feet. When tested by the Board of Trade Inspectors 1,830 tons were put on the 1,710 feet, but 4,000 tons might have been put on it without injuriously affecting the structure."

This address was delivered some time after the completion of the bridge, but, for the rest, he took very little part in the scientific discussions which occasionally arose during the construction of the bridge. He delighted in showing parties of his friends over the works, and in explaining to them the nature of the operations. Except the speech at the opening ceremony, and an article signed by himself and by Sir Benjamin Baker which appeared in the *Nineteenth Century* review, he left the work of popular exposition to his younger colleagues.

The responsibility, however, weighed much on his mind, and the zeal and energy which he threw into the work of general superintendence was unremitting. On March 23rd, 1884, he writes to his wife from Barcelona, where he had gone partly in search of health and partly on business.

"I sometimes analyse my motives in coming to Spain. The ostensible motive, of course, is to see Per's (his son's) work at the Lomo de Bas mines, as to which I have special duties as chairman, but I think the strongest of all is to endeavour to see the Forth Bridge finished, and to have a few more birthdays with you. You will say these are rather sad and morbid thoughts, and no doubt they are, but it is always the case with me when I have not sufficient occupation or absorbing interests."

"My chief consolation," he says in the same letter, "is the warm sun, and the consolation is rather the hope that it may diminish my cough than for the personal comfort, although that is something."

Later in the year, December 14th, he writes again from the Forth Bridge, to which he paid a fortnightly visit.

"I am very much disposed to come here after Christmas and make this my home for a month or two. I do not think the work will go on either rapidly or satisfactorily unless I do." Next day he writes again, "I am more and more convinced that I *must* come and live here for a time, if this work is to progress rapidly."

In October, 1886, he writes :—

"I have all the contractors here, and am hard at work organising a better progress with the works. They see I am terribly in earnest, and are carrying out all my suggestions with the energy I require. . . . The air is softer with the rain, and probably my cough will be less troublesome than it was with the cold of yesterday and the day before. I should be so pleased if I could safely and with reasonable comfort remain in England all the winter, and spend much of my time here. We must see."

We quoted earlier in this work a time-table of one of Sir John Fowler's working days. The same tireless energy still characterised his movements, as the following extract will show.

"FORTH BRIDGE, *May 29th*, 1887.

"I have almost settled to leave here very early on Thursday (3.30 a.m.) and go to Glen Mazeran, and leave there so early on Friday morning that I have an hour at Inverness with Dougal and Paterson on Highland railway matters, and go forward to Inverbroom by the twelve o'clock train from Inverness.

"On my way to Inverbroom I will call at Braemore House, and spend half an hour or an hour there whilst the carriage drives round to the square, also go through the garden before I go on to Inverbroom. I must leave Inverbroom on Thursday,

THE FORTH BRIDGE.

From the South-west, August 1, 1888.

To face page 311.

9th, to return here, which I can do on the same day, reaching Queensferry about eleven o'clock. This is a stiff programme, and involves early rising and long days, but if I am well, and my cough not very troublesome, I can manage it all, and it will be useful work."

Not a bad week's work for a man of seventy years of age! On November 17th, 1887, he writes again :—

" I am well this morning, and have just been over the works with Arrol, Falkiner, and Cooper. The progress of erection is now very satisfactory, and if we are all spared until next summer the operations will be more interesting than they have ever been. . . . I am very glad I came here now, as there are many matters in which my presence will be very useful. Yesterday the wind was so high that the work of erection was quite stopped. To-day is calm, and men are perched about in numerous places from 300 to 400 feet high, and apparently work as comfortably as if they were on the ground. . . ."

The opening ceremony took place under the auspices of the Prince of Wales, on March 4th, 1890. A gale of wind was blowing at the time, and the speech-making out-of-doors was cut very short. The last rivet of the bridge was driven by His Royal Highness, and the bridge was declared open. The company then sought the shelter of the banqueting-hall, erected at the Forth Bridge Station, where luncheon was served.

In his speech after lunch, the Prince announced that Her Majesty had been pleased to make Sir John Fowler and Mr. Thompson baronets, Mr. Baker a Knight Commander of the Order of St. George and St. Michael, and to confer on Mr. Arrol the honour of knighthood.

Sir John Fowler was called upon to reply. He said :—

" Your Royal Highness, Mr. Chairman, My Lords and gentlemen, I have to acknowledge this toast on my own part and on that of my colleague and partner Mr. Baker, whom I must learn as quickly as possible to call Sir Benjamin Baker, and on behalf of all our able engineering staff who have been associated with us in the design and construction of the Forth Bridge."

After referring to the class of critical pessimists who make it their business to declare that bold and novel undertakings like the Metropolitan Railway, the Suez Canal, and the Forth Bridge are impossibilities, he went on to say :—

"It is very curious to watch the manner of retreat of these prophets of failure when results prove they have been mistaken, and I could tell you some very curious stories connected with the Forth Bridge. But on this day I feel I can afford to be magnanimous, and I shall say nothing ill-natured about any of them—not even the astronomers. I am certain, however, that the astronomers are very sorry for themselves, because, since the failure of their predictions as to the Forth Bridge, cautious people are beginning to be a little doubtful about these small planets that they say they discover in the sky. Now, personally, I believe in astronomers, I believe even in their little planets, but I do not believe in astronomy being a safe guide for practical engineering. . . . I should like to designate this work as a British work. Scottish and English railway companies have found the capital. Aberdeen has found the granite. The greater part of the steel has come from Glasgow. The cement has come from the clay and chalk cliffs of the valley of the Thames, which, as you are aware, is in the neighbouring and very friendly kingdom of England, and part of the steel has come from ' dear little Wales.' The workmen have been chiefly Scottish. They are famous throughout the world as masons, and especially in granite; and I do not believe, and I profess myself to be a good judge, that a better piece of mason work was ever executed for any public work in this world." Then as to the æsthetic aspects of the work. "This, I confess, gave me a little concern, for I read rather strongly-worded letters in the newspapers, and altogether we seemed likely to get into hot water over the matter." However, a learned society in Edinburgh "called to its hall two famous artistic athletes—Mr. William Morris, as the vigorous attacker of the Forth Bridge on æsthetic grounds, and Mr. Benjamin Baker, the equally vigorous defender. The result, I believe, was that Mr. Baker and the bridge were entirely victorious." Then as to the durability of the bridge. "We have two materials in the Forth Bridge—granite and steel. . . . With Scottish granite connected with English cement we have durability and union of parts for at least a thousand years. . . . In regard to the steel, it can be deteriorated or decayed from two causes—vibration and oxidation. Vibration can only produce injurious action when the maximum strain to which it will be habitually exposed by use approaches to one-half of its ultimate strength; but as the Forth Bridge can never have as much as one-fourth of the ultimate strain, I think we may dismiss vibration as a source of injury. With regard to oxidation, that means gross neglect by those who have the bridge in charge; it means that the painting shall be so neglected that the atmosphere has direct access to the steel, and I won't do those who have the charge of this important and costly work the injustice of supposing such a contingency to be possible."

After grateful acknowledgment of the work done by his various colleagues — Mr. Baker, Mr. Allan Stewart, Mr. Phillips, Mr. Arrol, and Mr. Biggart— he then, last but not least, claimed their due meed of praise for those thousands of workmen — brave men, who for years, often in tempestuous weather, and at an elevation of one, two, three, or even four hundred feet above the waters of the Forth, did their hazardous work, and never knowingly scamped a rivet.

Sir John's allusion to the artistic aspects of the

question refers to an incident which is interesting and amusing.

Mr. William Morris, a great handicraftsman and man of genius, who unfortunately allowed his fine artistic sense to be obscured by a disordered political imagination, seemed to see in the Forth Bridge an embodiment of all he found amiss in the economy of our age.

"There never would be," he said, "an architecture in iron, every improvement in machinery being uglier and uglier, until they reach the supremest specimen of all ugliness—the Forth Bridge."

In reply to this tirade Sir B. Baker, lecturing the following evening (November 27th, 1889) at the Edinburgh Literary Institute, not unnaturally expressed a doubt

"if Mr. Morris had the faintest knowledge of the duties which the great structure had to perform, and he could not judge of the impression which it made on the minds of those who, having that knowledge, could appreciate the direction of the lines of stress and the fitness of the several members to resist the forces. Probably Mr. Morris would judge the beauty of a design from the same standpoint, whether it was for a bridge a mile long, or for a silver chimney ornament. It was impossible for anyone to pronounce authoritatively on the beauty of an object without knowing its functions. The marble columns of the Parthenon were beautiful where they stood, but if they took one and bored a hole through its axis and used it as a funnel of an Atlantic liner it would, to his mind, cease to be beautiful, but of course Mr. Morris might think otherwise."

"He (Sir B. Baker) had been asked why the under side of the bridge had not been made a true arc, instead of polygonal

METHOD OF ERECTION.

Showing the polygonal character of the cantilever span (May 24, 1889).

To face page 315.

in form, and his reply was that to have made it so would have materialised a falsehood. The Forth Bridge was not an arch, and it said so for itself. No one would admire bent columns in an architectural façade, or a beam tricked out to look like an arch; but that was really to what the suggestion of his artistic friends amounted, though they did not see it, being ignorant of the principles on which the Forth Bridge was constructed. Critics must first study the work to be done both by the piers, and by the superstructure, and also the materials employed, before they are capable of settling whether it is beautiful or ugly. It would, he added, be a ludicrous error to suppose that Sir John Fowler and he had neglected to consider the design from the artistic point of view. They did so from the very first. An arched form was admittedly graceful, and they had approximated their bridge to that form as closely as they could without suggesting false constructions and shams. They made the compression members strong tubes, and the tension members light lattice work, so that to any intelligent eye the nature of the stresses and the sufficiency of the members of the structure to resist them were emphasised at all points. It would have been futile to attempt to ornament the great cantilevers, and so, to keep the whole work in harmony, they studiously avoided any attempt at ornamentation of the piers, and people would search in vain even for a moulded capping, or cornice throughout the whole work. The object had been so to arrange the leading lines of the structure as to convey an idea of strength and stability. This, in such a structure, seemed to be at once the truest and highest art."

In reference to this interesting question of the merits or demerits of the bridge from an artistic point of view, we may fitly conclude our account of this great work by quoting from a letter of congratulation written to Sir John Fowler, when the bridge was completed, by a competent critic, Mr. Alfred Waterhouse, R.A. :—

"20, NEW CAVENDISH STREET,
PORTLAND PLACE, W.,
"*December 5th*, 1889.

"DEAR SIR JOHN FOWLER,—I was very much obliged to you for so kindly enabling me to visit the Forth Bridge in so satisfactory a manner. Your note was the means of my introduction to Mr. Baker, who happened to be on the spot, and who was very good to me as introduced by you.

"I expect you are tired of congratulations, but I cannot help saying how overawed I was by the work. It hardly looks human. One feature especially delights me—the absence of all ornament. Any architectural detail borrowed from any style would have been out of place in such a work. As it is, the bridge is a style unto itself.

"The simple directness of purpose with which it does its work is splendid, and invests your vast monument with a kind of beauty of its own, differing though it certainly does from all other beautiful things I have ever seen.

"Believe me,
"Yours very truly,
"A. WATERHOUSE."

It would be endless and unnecessary to record the congratulations and public laudatory notices which followed the completion of the great bridge. The Sovereign expressed her approval by conferring titles on the engineers; the University of Edinburgh honoured itself by giving the degree of Doctor of Laws; but perhaps the most remarkable acknowledgment of all was the bestowal of the Prix Poncelet on Sir John Fowler and Sir Benjamin Baker by the Institute of France, a gratifying and impartial tribute to the fact that scientific generosity and merit rise above the narrow limitations of nationality and creed. At this date there was no other precedent for the bestowal of this international honour on an Englishman, except in the noteworthy instance of Lord Kelvin, who had already received the same honourable distinction.

Those, however, who have seen the bridge cannot but feel that there is a certain incongruity in seeking to increase, by such means, the fame and honour of its builders. The most colossal structure of the foremost industrial nation of the world, standing as it does in a place where the natural scenery sets off its superb proportions, is itself the most appropriate memorial of its makers.

CHAPTER XII.

THE ENGINEER AT HOME

IT is, of course, in his professional capacity that Sir John Fowler becomes an outstanding figure in the engineering world of the second half of the nineteenth century, and we have endeavoured to set out the conjuncture of character and opportunity from which his public career proceeded.

When we speak of a man's private life we are ordinarily supposed to refer to those leisure hours which he spends away from his business. The formative influence in Sir John Fowler's life and character, however, was undoubtedly his business experience. To the incidents of his daily work, his energy, punctuality, attention to detail, and that probity of character which is at once controlling and exacting, both to himself and others, were readily applicable, and have already been noticed. It remains to consider how a character so profoundly influenced by the necessities of an industrial age appeared in the more spontaneous and impulsive atmosphere of home-life.

Fowler was naturally a man of warm affections, and, though in early life he yielded to it very little, with a deep strain of sentiment in his disposition. He was a strong man, but he was also a very good-natured man, and the fact made him on the whole a kindly

minister of those economic laws which some find so
hard and unrelenting. He was, in this way, a popular
man in his own profession and in society. He showed
a generous desire to obtain public recognition of the
work of other members of the profession, and expression
of his opinion had weight with those who had the
responsibility of dispensing public honours. Among
other notable engineering works of the last few years,
he was much interested in the Manchester Ship Canal,
and on the passage of the Bill through Parliament, and
on the completion of the work, he wrote warm letters of
congratulation to the engineer, Mr. (now Sir) E. Leader
Williams. " I am sure," he said, " all your friends will
be pleased and proud, as I am, at the result of your
long perseverance against formidable odds." The compliment was duly appreciated. Sir E. Leader Williams
writes : " Sir John always took an interest in the
success of other engineers in carrying out large works.
He knew by experience the anxiety and worry that
attends the prosecution of such enterprises, and his
kind words and letters were prized by his professional
brethren."

A man's private and public life is not, though we
sometimes talk as if it were, divided into two water-
tight compartments, and Sir John Fowler's case was no
exception, for just as his genial and kindly nature
followed him into his professional life, so occasionally the
somewhat arbitrary habit of thought, acquired by the
practice of command and the achievement of success,
coloured the judgments and actions of his private life.
This must be the explanation of a certain controversial
keenness and love of mastery which, especially in later
years, took possession of him. In professional matters

this tendency was, as we have remarked, well under control; when necessary he knew how to yield, but naturally, in his expression of political opinions, in the management of his family and affairs, no such imperative occasions for concession arose. Happily in such matters, as we have seen in the important question of standing for Parliament, he sought and was much guided by the advice of his wife. Her womanly tact had a refining and moderating influence on a temperament that was naturally imperious.

To complete our memorial of Sir John Fowler's career, some sketch must here be attempted of his home-life in London and at Braemore.

Up to 1867 he continued to live at 2, Queen Square Place. Here, as elsewhere, he enjoyed keeping open house for his friends. Professional men then lived, if the term may be allowed, on greater terms of intimacy with their business. The ladies of the family occasionally penetrated into offices which were under the same roof as the dwelling. During the making of the Metropolitan Railway, when much of the work had to be done in the midnight hours, entertainment used to be provided for the staff by the hospitable chief. In the early sixties, and especially during his presidency of the Institution, he had every Tuesday night a small dinner-party, to which a select circle of engineering friends had a standing invitation. Later in the evening the party adjourned to the weekly meeting of the Institution. After a day's labour of opposing one another in the parliamentary committee-rooms over the way, there would assemble round his hospitable board the younger Stephenson, Locke, Brunel, Bidder, Scott-Russell, Rendel, and Hawkshaw.

No record, alas! has been kept of these interesting gatherings, though with Lady Fowler, who was always, and with Mr. Baldry, who was sometimes, present, they live as memories of pleasant and brilliant talk.

In 1867 the Fowlers removed to Thornwood Lodge, a house with a large garden and grounds on Campden Hill, Kensington. Mr. Fowler had become acquainted with the amenities of this charming *rus in urbe* while the house was in the temporary occupation of his friend Mr. (now Sir) William Vernon Harcourt. Here he had his London headquarters for the remainder of his life.

He was fond of society, and went everywhere. Society, too, liked him, and he was accepted socially as, in an informal way, the representative of his great profession. He enjoyed being of service to his friends in little matters of domestic engineering. In this way he took great pleasure and interest in designing and superintending the making of the boys' bathing-place at Harrow, where his sons were at school. He also was consulted about the levelling of the cricket ground.

He had been a keen cricketer in his youth, and as a member of the M.C.C. he was a familiar figure at Lord's, where he seldom failed to attend at the great matches of the year.

He had a Yorkshireman's love of a good horse, and, when in London, he rarely missed his morning ride in the park. He was a man of great bodily strength, and he used to say that till he was sixty he did not know what it was to feel physical fatigue. He was a light sleeper and woke early. He rose every morning at six, made himself a cup of tea, and wrote his letters between seven and nine.

Y

We have already noted his first purchase of land at Glen Mazeran, Inverness-shire. He soon became ambitious of something on a larger scale, and in 1865 he purchased the estate of Braemore, in Ross-shire, and in 1867 the adjoining estate of Inverbroom.

In a letter to his father dated July 15th, 1867, he gives the following account of his new purchase :—

"The chief works which I am carrying out are building the new house and the garden and outbuildings, roads, etc., belonging to it, improvement of the arable land, and plantations.

"The house is a serious undertaking, both as regards the carrying out of such a large work in such a wild, uninhabited part of the country and as regards its cost. I think it is so far advanced that no doubt remains we shall be able to cover it in before the end of October, and have it ready for occupation next season. The bulk of the stone for the stonework is obtained from a quarry about a quarter of a mile from the house. It is a rock very common in the north of Scotland called gneiss. It is blue in colour, consisting of felspar, quartz, mica, and hornblende, and therefore in its component parts not unlike granite, but it appears to have been subjected to a less degree of heat than granite, as it retains its schistose or stratified character. It is wonderfully durable, and makes excellent work for the plain part of the walls, but it cannot be worked into anything requiring an acies or edge, and therefore we have window-heads, sills, plinths, etc., brought from Glasgow, where quarries supplying the finest compact and durable sandstone abound. It is brought by sea to the head of Loch Broom, and then carted six miles to the house. The contrast of the sandstone with the deep blue gneiss produces a good effect, and I expect the green Westmoreland slates will also accord well as to colour.

"The people in the country have been very much astonished by my building the house 700 feet above the sea, instead of following the old Scotch practice of burying it in the lowest place I could find, but now they see what is likely to be

THE HOUSE OF BRAEMORE.
Looking towards Loch Broom.

To face page 322.

realised, and the plantations springing up about it, the new idea seems to find favour.

"All the plantations are promising well, and I believe in this climate with a southern and western aspect, the trees will make rapid progress after two or three years. Road making is easy, as the materials are good, and my Glen Mazeran tenant, Clunas, is especially good at road making."

Then follows a statement of his intentions with regard to the farming of the lower part of the strath, and the letter ends in a strain of moralising very unusual with the writer :—

"The garden promises to be a great success, and next year we may expect to be well supplied with all kinds of vegetables and fruit. . . . You will see by the date of this letter that it is written on my birthday, and that on this day I am fifty years old. To you this will almost appear the age of youth, but to me it appears the commencement of old age. I scarcely know why, but age seems to have stolen upon me unawares, and taken me by a curious surprise. I sometimes think it is because for many years my life was so fully occupied with daily work that I had no time for retrospective and prospective thought, but probably it is the case with all.

"For any measure of worldly success I may have had I have to thank you, my dear father, for the early care you took of my education, and the choice of my career, and the examples you have always presented to me and to your children of truthfulness and strict integrity. May you yet be spared a little longer with your clear intellect and great intelligence, and as free from pain and every discomfort as is possible for you. . . ."

Together the estates of Braemore and Inverbroom contain some 40,000 acres. For the most part the land was wild heather and rock, destitute of trees except in the lower parts of the strath, where, bordering on the rivers, there was a small stretch of arable

land.   Mr. Fowler set to work with characteristic
energy to develop the resources and amenities of this
beautiful tract of country.

First he determined to have a view from his house.
He chose, therefore, a site on the spur of the mountains
that flank the valley of the River Broom.   It was a
daring adventure, even for a great engineer, and friends
were not wanting to hint that the aerial mansion would
be uninhabitable.   The result, however, has fully
justified the effort.   The wood has grown up, and the
house, being built fair and square, is warm and comfortable even in the wildest weather.   The stables and
the gardens are situated directly below the mansion
house, on the level of the high-road.   A winding road
of some miles in length and of easy gradients leads
from the high-road to the house, but a short, if somewhat precipitous, path runs straight down from the
house to the stables and gardens.   The streams of the
mountain-side have been gathered into a loch some way
above the house, and utilised to furnish power for
electric light and other purposes.

The road which leads from Garve Station, on the
Dingwall and Skye Railway, to Braemore runs for miles
through a desolate moor.   There are traces, however,
everywhere, of a primæval forest, a subject about which
Sir John Fowler was never weary of speculating.

Shortly before reaching the watershed between the
east and west of Scotland, we come on the Braemore
property, and as we pass the watershed and reach the
shores of Loch Drome, we see the first signs of Sir John
Fowler's experiment in re-afforesting the country.   The
situation of the land at this point, though exposed
to the milder breezes of the west, has proved too

CORRIE-HALLOCH, BRAEMORE.

To face page 325.

high and wind-swept, and the plantations here have
not been so successful as lower down the valley, where
some 1,200 acres, which from time immemorial have
been bare mountain, are now successfully covered with
Scotch fir, larch, spruce, and hardwood. Altogether
some 9,000,000 of trees have been planted along the
strath. The contrast between the bare and dreary
country which the traveller passes on the ascent from
Garve, and the sylvan landscape into which he now
descends, is most enchanting.

The views from the house extend northwards down
the valley of the Broom to the sea loch of that name,
and towards the seaport of Ullapool, while to the
west the prospect stretches into the fastnesses of the
forest. In the immediate neighbourhood of the house
the zealous care of Sir John and Lady Fowler has
collected every variety of heath and of mountain tree
and shrub. Most wisely, even in the gardens, the
attempt has been abandoned to acclimatise exotics. No
effort, on the other hand, has been spared to realise,
under cultivation, the unsurpassable beauty of a Highland mountain side.

The River Broom for more than a mile of its course
runs in a precipitous gully, at places some 200 and
300 feet in depth. Across the river it has been the
congenial amusement of the engineer's holiday to
construct three "very handsome light iron bridges,"
one for carrying a roadway across into the forest, the
others footbridges to allow his visitors to see the
magnificent waterfall of Corrie Halloch and the wild
gorges of the river. From a point not far below
the house the river issues out of a succession of rocky
gullies to the open strath, and about four miles lower

down flows into the tidal water of Loch Broom. Here, on the river also, every device of which an engineer can dream has been called into use to ornament and improve the property. Embankments have been made, and the adjoining land has been trenched and drained and converted into pasture and arable field. Loch Drome, the sheet of water which is passed at the head of the valley on the road from Garve, has been increased from 37 acres to about double that extent, by means of an embankment at the lower end. This operation has much improved the fishing in the loch. A small flood in the river also can be assisted by opening the sluices in this embankment, and the device goes some way to realise the fisherman's paradise, in which he can control his own "spates." It has been stated in print that this artificial flood has been efficacious for purposes of salmon fishing, but on this head we regret to find that scepticism still prevails.

The Scottish Highlander, whether of high or low degree, is not much inclined to welcome the Sassenach settler with effusion, but Sir John Fowler and his family very early succeeded in overcoming this clannish exclusiveness. His neighbours magnanimously overlooked his misfortune in not having been born a Scotsman. At a dinner of the Scottish Corporation, he told his hearers that he had done his best to be a Scotsman. His worldly goods were mostly in Scotland. For forty years of his life he had never missed spending his autumns in Scotland. His eldest son resided there. His grandchildren were born there, and wherever he went his heart was in the Highlands. He had inquired as to the objects of the charity in the support of which they were feasting, and had been glad to learn that neither religion nor nationality was a bar to membership, but he observed that only Scotsmen were qualified to receive—a very good rule —and he wished the Scottish Corporation all success.

Scotsmen have a great admiration for the man of solid achievement, and they like their humour to be served to them dry. Sir John Fowler succeeded admirably in catching the tone of his adopted country. As he passed, he had always a friendly greeting and a pleasant jest for his humble neighbours. He built good cottages, and was a kindly and improving landlord. His practical knowledge of farming and building and generally of estate work won him the respect of his tenants. "He knew," one of them remarked, "the prices and quality of the material used in all such operations as well as any of us. No one could get the better of him in a bargain." The usual ignorance of the cockney shooting tenant on such topics might lead to profit, but not to popularity. The shrewd, capable, and liberal landlord was a more congenial personage. His knowledge and keen business habits seconded and encouraged the industry of his tenants. In a word, he recognised that there was a business side to the duties of a landlord, and very faithfully he performed his share of the contract.

Visitors would occasionally appear from yachts, storm-staid on the coast. At one time it was Mr. W. H. Smith and at another Lord Rosebery who came over from Ullapool to see the sylvan fairyland which the laird of Braemore had created on the barren mountain side. But for the most part, in a sparsely populated region like Ross-shire, the occupants of the great houses are necessarily dependent for society on their

own visitors. The hospitality of Braemore was among
Sir John's principal pleasures. There came to Braemore, year after year, a long succession of visitors,
and for them the greatest variety of entertainment
was provided. The company was not exclusively
composed of sportsmen. Indeed, a confidential talk
with McHardy, who for thirty years was head-keeper
and stalker, and trusted friend of the family, has left
the impression that the sportsmanlike prowess of the
visitors varied inversely to their distinction in other
fields of fame. A visitors' book with ample margin
contains a record of distinguished names. Great
artists—Landseer and Millais—have adorned the pages
with charming sketches in reminiscence of pleasant
days spent in the forest or on the loch. High
ecclesiastics and statesmen contributed to the volume
appropriate sentiments in prose and verse.

" When the dull, dreary session is over, and patriots twaddle no
more,
How blithely I breathe the brave breezes which blow round the
braes of Braemore !
Though the Broom like our Gladstone meanders, or foams down
with froth in a spate,
Though the stalker, like Dizzy in ambush, for his prey is aye
lying in wait,
Yet here may we cast away care," &c. &c.

So sings a distinguished statesman who is not generally associated in the minds of his countrymen with
the pursuit of the muses.

For the fishermen there was a stretch of four miles
of salmon river, and many lochs in which trout
abounded. On the low ground of the forest a good
day's grouse shooting was to be had, and some four miles

THE RECREATIONS OF A STATESMAN.

Sir William V. Harcourt, by Sir J. E. Millais, P.R.A.

(From the Visitors' Book at Braemore).

To face page 328.

STAGS AND HINDS. By Sir Edwin Landseer, R.A.
(From the Visitors' Book at Braemore).

To face page 328.

away, at the head of Loch Broom, Sir John's yacht
was moored, and especially during his later years it
was his delight to organise a water picnic for a party
of congenial friends.

Sir John at different times owned three yachts,
each of them named *The Southern Cross*, after the
constellation which so delighted him during his stay
in Egypt. His first yacht was built for him in 1878.
Some years later he bought a 350-ton yacht, in which
he visited the Mediterranean and cruised among the
islands of the Greek archipelago. This was in later
years replaced by a smaller and handier craft, more
suitable for the work required of it in the bays and
narrow channels of the Scottish coast. He was elected
a member of the Royal Yacht Squadron, a social distinction much prized by those who, for their pleasure,
go down to the sea in ships.

Last, but of course not least, there was the deer
forest—that sport of kings. Sir John himself was a
keen deer stalker; indeed, to him all other sports
except salmon fishing were matters indifferent. In
later years he rarely fired a gun at a grouse or threw
a fly for a trout, and he took no interest in the big
battues, common in south country sport. Notwithstanding this somewhat exclusive taste, Sir John was
emphatically what is called "a good sportsman," a
phrase of talismanic import to the initiated. There
are presumably varieties of "good sportsmen." There
is the man who, whether born in the palace or in
the cottage, has obviously been intended by Providence for a gamekeeper, who lives on affectionate
terms with ferrets, and who, in default of higher
quarry, will take part with enthusiasm in a rat hunt.

Sport with Sir John was a less absorbing interest; he liked only the best of it, and he took a pride in seeing that everything was properly done.

Sport, however, was not his sole interest at Braemore. He was always very fond of getting experts to talk to him on their own subjects. Geological rambles with Sir Roderick Murchison were red-letter days to him.

The following entry in the Visitors' Book at Braemore shows that the interest of these rambles was highly valued by that great geologist himself:—

"Adieu, Braemore!" he writes, "where the cordial reception of the kind host and hostess have made an indelible impression on the heart of the old Silurian!

"Forty-two years have elapsed since, when in company with Professor Sedgwick, I hammered the rocks at Ullapool, and now, by the active assistance of Mr. Fowler and the aid of his handy steam yacht, I have been enabled to place all the great rock formations which are exposed on the shores of Loch Broom in their true order of age and succession, from my fundamental gneiss (*hodie Laurentian*), through the grand massive Cambrian rocks of Ben More, the lower Silurian quartz rocks and limestone of Ullapool, up to the overlying gneiss of Braemore (metamorphic lower Silurian), on which the mansion stands, and from which, looking northwards, the spectator commands in one unrivalled view all this glorious geological series.—August 21st, 1869."

His lifelong friendship with Sir Richard Owen gave him similar occasions of discussing the conditions of prehistoric life. One excursion which the two friends took together resulted in a trophy of national interest.

"You remember," writes Professor Owen, "our excursion to Sandside! The skeleton of that whale now forms the chief feature in the grand hall of the new museum (the Natural History Museum at South Kensington), and I am daily gratified by the interest with which it is inspected."

To Braemore also came the extremes of political parties — Sir William Harcourt, whose acquaintance and friendship had been made in parliamentary committee-rooms, Earl Cairns, the great Conservative Lord Chancellor, the veteran statesman Lord Cranbrook, and other notabilities, too long a list to narrate.

On Friday, August 11th, 1871, there is a pathetic signature in blue indelible pencil, with regard to which Sir Richard Owen writes in his journal:—

"I am now the only guest. Lady Ashburton and T. Carlyle drove over and took tea with us. . . . He is much emaciated, can digest but little, and hardly gets any sleep. He was most friendly, and, I thought, took his last leave of me at parting. . . . He painfully with a pencil put his name in the Visitors' Book."

Professional friends, his partners Sir Benjamin Baker and Mr. Baldry, usually gave a few days of their holidays to a visit at Braemore. His old antagonist — now Field-Marshal — Sir Lintorn Simmons, was also a regular summer guest. Here, too, came Prince Hassan, son of Ismail the Khedive, at that time a student at the University of Oxford. Professor Owen records in his diary under date, August 20th, 1871, that the Prince

"has quite fallen into English ways, and speaks English perfectly. On Monday he made his first stalk on the mountains, and was so excited by the thought of it that he threw all the cushions about in the drawing-room!"

Archbishops and bishops, and lesser ecclesiastical dignitaries such as deans and archdeacons, as well as eminent laymen from many varied walks of life, abounded, all eager to please and to be pleased, for so the kindly wizard of the mountain ordered.

When the labours of the day were over, the party met at dinner in the highest good humour. It was holiday time with all. The railway station, that portal of dull care, was more than 20 miles distant, the scenery was enchanting, and the day's sport had been invigorating; everything was propitious for the relaxation of frank and cheerful talk. Each member of the company had something worth hearing to say. The Field-Marshal had anecdotes of the secret history of the Crimean War, or of Lord Beaconsfield at the Berlin Conference. It was an hour at which even an archbishop might be indiscreet. All this was a sincere pleasure to the genial host.

The late Archbishop of Canterbury, to whom Sir John's son Montague was for a time domestic chaplain and assistant secretary, was a frequent and welcome guest. The Archbishop had at first excused himself from accepting an invitation to Braemore on the ground that he liked to spend his holiday with his family. Sir John, however, was not to be denied, and as the Archbishop and Mrs. Benson would not leave their children, he invited all the children to accompany them. So it came about that the whole family on more than one occasion passed several happy weeks at Braemore. Mr. Arthur Benson, the archbishop's son and biographer, has obligingly sent the following pleasing and authentic picture of the life at Braemore.

"Sir John Fowler, as I first remember him, was a strongly-built man, with large, rough-hewn features. His whole face spoke of work—hard, practical work. He evidently never gave his personal appearance a thought. He dressed in loosely-fitting clothes; he was becoming bald, and his strong hair, which he wore long, stood out stiffly from the back of his

head; he was clean-shaven, with the exception of rough
whiskers and a small beard under his chin; the somewhat
downright expression of his full, mobile lips, often pursed
in thought, his big nose, his full-blooded complexion, was
humanised by the genial and kindly glance of his light-
coloured eyes. He would have despised any elegance of
motion, hand, or demeanour, and his brusque gestures and
walk would have been almost clumsy but for the dignity of
strength which characterised all he did. At the same time, he
was a man of quick and sensitive emotions, which contrasted
strongly with his bluff exterior. I have often seen his eyes
fill suddenly with tears, which he would brush away with his
hand, and his voice falter when he talked of anyone, and there
were many, who were dear to him. I remember that when
I first saw him in Scotland, in an old brown shooting-coat, he
seemed to me at once to move with greater simplicity and ease
than in formal London garments. In the evenings in Scotland
he wore a velvet suit with knickerbockers and purple stockings,
in which I used to fancy he looked like a benevolent artist.
But this dress, which might have seemed *posé* in some men, only
appeared in his case natural and without calculated effect. . . .

"Sir John Fowler's gift of hospitality was very great. To
hit the exact mean in entertaining guests requires a natural
talent. It is possible to leave them so much alone that they
feel neglected, or to fill the day with a ceaseless round of
engagements, until weariness results. At Braemore everything
was perfectly organised; there was no sense of proprietorship
about Sir John: he sate like a guest among his guests; and yet
every day was laid out. As soon as breakfast was over he
would say what he proposed: he would tell my father that
he would leave him free till luncheon, and then he would take
him for a ride in the forest to see the deer at home. Then came
the turn of each of us. Lady Fowler, of whose delicate and un-
obtrusive kindliness I may not here further speak, would be at my
mother's disposal. Then my sisters were provided for; gillies
placed at the disposal of myself and my brothers, and I do not
recollect a single day upon which Sir John had not some plan

for my youngest brother, then a schoolboy of fourteen. He certainly delighted in making these arrangements, but they were never forced upon anyone, and if a preference was expressed, matters were at once skilfully accommodated. I do not remember that he took much physical exercise himself. I remember his occasionally stalking a deer, and once or twice he took a salmon-rod to the river. He was a successful sportsman, but he knew that my father had little sympathy for any kind of sport, and a deep-rooted objection to depriving any creature of the inestimable privilege of life, and so, though the younger members of the party were amply provided with fishing and stalking, yet the fact was never obtruded.

"I imagine that he must have done a great deal of professional work in the course of the day, but it was all kept very much out of sight. He was fond of sending and receiving long telegrams; more than once I remember some interesting news arriving from town on telegrams covering four or five sheets. . . .

"He had a strong vein of humour, and I can remember his relating with high relish how he had been badgered in some important case before a Parliamentary Committee by an eminent Q.C., who asked him all kinds of ingenious questions, which, though bearing little on the matter in hand, were yet difficult to answer satisfactorily. 'Now, Mr. Fowler,' said the counsel, 'let me ask you this. Supposing two of the largest vessels in Her Majesty's fleet were to collide beneath this structure of yours, what would be the result? Let us have a candid answer.' 'I imagine,' said Sir John, 'that both the captains would be dismissed from the service for gross incompetence.' The Forth Bridge was then in course of construction, and though Sir John must have been tired enough of being questioned about it, he was always ready to give some of the astounding figures connected with the materials used in the construction. One day at luncheon he explained the principle of cantilevers with a piece of paper and a pencil, illustrating it with a cork and a pair of forks, with a precision and brevity which astonished me. I can only say that, with no mechanical predilections, I have never forgotten it. . . .

SIR JOHN FOWLER AND HIS ELDEST GRANDSON.

To face page 335.

"To my mother on another occasion he spoke of the delight he took in managing men. 'It is the greatest pleasure in the world,' he said, 'to have a thoroughly capable man to deal with, who is at the same time very difficult to manage, so that you have to try all the flies in your book. I never allow myself to be beaten, and there is nothing in the world I enjoy more.'"

The secret of all true courtesy and hospitality is the desire to please and to be pleased. At Braemore these conditions were thoroughly fulfilled. The result was an all-pervading atmosphere of genial enjoyment which none could resist.

Sir John had, we are informed by a competent observer, a curious habit of taking stock of strangers. It consisted of "drawing" them on their special subjects. He was at pains, at first at all events, to conceal any imperfections there might be in his own knowledge of the question discussed. The problem in his mind seemed to be, "Is this man an impostor, or is he not?" The ordeal of examination, of which the victim was probably quite unconscious, was often somewhat searching, but at length the *dossier* of the examined was silently put away for future use in some mental pigeon-hole. Nothing in most cases came of this appraisal of men. His verdicts, however, were sometimes communicated to those in his confidence, and were seldom far from the truth. It was an object with him to know who were the leading intellects in every sphere of human activity, and his position in society gave him frequent opportunity of indulging this foible. Like all men of kindly disposition, he was devoted to children. Yet even in his intercourse with them, he showed the same keen desire to discover talent and to watch its development. An almost exaggerated

respect for intellectual ability was a marked characteristic of his mind.

We have left for the last chapter all reference to
Sir John's political opinions. We have merely noted
that he was a strong Conservative. In 1879, at an age
when a prominent parliamentary career was no longer
probable, he was selected to contest Tewkesbury in that
interest. Lord Beaconsfield had recently returned from
Berlin bringing peace with honour, and the next election
was to be fought in the light of that event. A speech
which he delivered early in January to the electors
contains a statement of his political creed.

He was keenly in favour of a forward foreign policy
in Egypt and on the Indian frontier, and much opposed
to the masterly inactivity which was being recommended in other quarters. He would do all in his
power to improve the health and education and comfort
of the people, but he was against revolutionary changes,
disestablishment, and, to use his own phrase, "depriving
a poor man of his beer."

Like all men of his generation, he was a staunch
believer in self-help. Though, as already noticed in
quotations from his article on railway accidents, he was
very far removed from the doctrinaire opponents of
Government regulation, he yet retained that attitude
of suspicion and dislike towards Government interference which was characteristic of the Liberalism of
an earlier period. It cannot, however, be said that he
was altogether consistent in his political principles.
In the short period of his Hallamshire candidature he
seems to have been not disinclined to dally with those
who favoured that very serious form of Government
interference known as fair trade or protection. His turn

of mind was practical rather than theoretic, and we are informed, that from practical considerations he accepted as inevitable for England the policy of Free Trade. Indeed, except on the subject of Egypt, we doubt if he really was much interested in politics. His orderly mind hated the confusion and anarchy caused by Mr. Gladstone's abnegation of responsibility for the well-being and progress of the Egyptian Soudan. The bungling, which in practical matters is apt to attend the indecision of the philosophic temperament, was abhorrent to him. The Titan might be weary of the burden of empire, but in John Fowler's view he must not decline his allotted task. His professional connection with Egypt gave additional emphasis to this criticism. For the rest, it appears to us, that he adopted the policy of the party to which he belonged very much in the same spirit as he accepted the verdict of experts on scientific problems with which he was not personally conversant. His vehemence of expression was not so much a sign of enthusiasm as of the latent energy which characterised all he did. To the theoretical aspect of politics he does not seem to have given much attention.

The following letters addressed to the *Times* on the subject of the rescue of Gordon seem worthy of being put on record. They were at the time an original and weighty contribution to political controversy.

"THE SITUATION IN EGYPT.

"*To the Editor of ' The Times,' April 30th, 1884.*

"SIR,—I have hitherto resisted great pressure from many friends deeply interested in Egyptian affairs, and have refrained from any public expression of opinion on the great problem which has long agitated all classes in England; but the position

z

is now so critical that I feel a duty is imposed upon all those who, like myself, have had long and intimate knowledge of Egypt, to speak plainly their views without reserve or delay. It would be idle and useless to enumerate the incidents which have led to the present deplorable state of things, although at some future time, less urgent than the present, a valuable lesson may be learnt from them, and therefore I will at once proceed to set forth in a few words the warning I desire to convey. Throughout England a cry has arisen that Gordon must be rescued at any cost, if possible, from the perilous position into which his noble nature has led him, and it is now proposed to make the attempt. I venture no opinion on the mode of making this attempt, but I do emphatically say, from my knowledge of the people and the country, that before any further step is taken involving English blood and treasure, England must distinctly and unequivocally state her intentions and future policy with regard to Egypt. I have every reason to believe that if England would even now plainly declare throughout Egypt that circumstances had imposed upon her the responsibility of the security and good government of the country, and that she had determined to assume that responsibility, all serious difficulties would soon vanish and danger to Gordon be averted.

"No mistake can be greater, in my opinion, than to suppose that such a course would involve either greater expenditure or greater military exertion than the policy of the past, which, if continued, would, I fear, lead to continued humiliation and certain failure. On the contrary, I believe that if we promptly declared our intention to assume the responsibility which everyone personally acquainted with the country must know to be inevitable, we should find an expedition up the Nile would be received with welcome and not resisted.

"The Egyptian people are easily governed if treated with kindness and justice, as they would be under an English administration, and all Europeans, without exception, would profit by the establishment of such a stable government if felt and known to be continuous and permanent, which is an essential condition of success. I refrain from any suggestions respecting details of administration or improvement and development of the country, which I have had unusual opportunities of studying, because I desire to limit myself at this time to the expression of an earnest hope that no further steps will be taken by England in Egypt, either defensive or offensive, until a distinct policy has been pronounced, so that the people in every part may know the Government which they have to obey, and by which they will be protected.

> " I am, Sir, your obedient servant,
> " JOHN FOWLER.

"No. 2, QUEEN SQUARE PLACE, WESTMINSTER."

" *To the Editor of ' The Times,' May 7th,* 1884.

" SIR,—I make no apology for presuming to occupy a little more of your space on Egyptian affairs. The rapidly advancing and already impending dangers in Egypt are not, I fear, generally understood in England, to their full extent, and, strange to say, they seem to be the least recognised in any sufficient degree in the House of Commons.

" Official and semi-official reports of the last few days announce a startling increase of excitement and of insolent defiance of authority in Upper Egypt, and private letters from trustworthy sources confirm these reports, and consider them to understate the real facts of the case.

"A correspondent who has been more than twenty years in Egypt, with unusual opportunities of knowing the feelings of the natives, and for whose perfect reliability I can personally vouch, writes as follows :—

" 'We have been going from bad to worse, and now we are in a state of anarchy. Robberies are being committed in the provincial towns in the face of day, because there is nothing to prevent it. The old police were bad, and have been abolished, but they were of some use, and nothing has been put in their place. Burglaries and shooting from behind hedges is increasing and will increase, and lately an attempt by the authorities to suppress a mischievous newspaper was met by defiance; and so things move on. Meanwhile the poor country suffers.'

"Yes, that is just it, 'and so things move on,' and so they must, so long as we remain in the position of temporary 'caretakers'; for we are really nothing else, and we are doing great and daily injury to Egypt and ourselves by a continuance in this anomalous position.

"The long, interesting letter of Sir Samuel Baker in *The Times* of the 30th of April, sets forth in great detail, and with the advantage of full local knowledge, the measures which he considers necessary for military success and the establishment of English authority. It is not probable that the time, men, and money required for his proposal will be available, but the proposal itself is a remarkable proof of the consequences which we have already incurred by neglected duties.

"If we discard as impracticable, which I think we may, such proposals as abandoning Egypt to take care of itself, or return to dual, or still more divided, control, and Sir Samuel Baker's scheme, there remains only the policy which alone was the possible one after we once went to Egypt to put down rebellion—viz. the acceptance of full responsibility of security to life and property, and the good government of the people.

"If we were not prepared for this we had no business to go to Egypt alone as we did. With regard to withdrawal, even that miserable step would be better than a continuance of our present aimless, do-nothing policy, and probably we could withdraw now, while if we wait longer we may meet the fate we have recklessly and persistently tempted.

"No such disgrace, however, is necessary, nor what is called 'the abandonment of the Soudan,' which means nothing less than leaving to anarchy and slavery a vast district capable of cultivation and of gradual civilisation, and almost compelling it to remain a perpetual menace to Egypt proper. A few years ago I had numerous engineers and surveyors engaged during two seasons in Darfur, as far as the capital El Fashr, in various parts of Kordofan, and also to Shendy and Khartoum, and during that long period, and over many hundreds of miles of country, no trouble whatever was experienced with the natives. They were well aware that our objects were peaceable and intended for their good, and that we were under the protection of a real Power at Cairo, which they recognised and respected.

"During this time I was in constant communication with Gordon Pasha, both in Cairo and when he was in Khartoum, and in various parts of Kordofan and Darfur, and certainly nothing was further from his views at that time than 'the abandonment of the Soudan.' On the contrary, he was a warm advocate for its development and peaceable settlement, and explained to me his plans for securing permanent frontiers on the side of Abyssinia and also to the south and west. These plans and his general views appeared to me admirable, and calculated to lead to the total extinction of slavery, but they were based, or they would have been futile, on a continuance of regular government and adequate authority.

"At that time neither Egyptian nor English prestige had been lost, and therefore authority was easily maintained. We have foolishly thrown away and temporarily lost this precious prestige, and there is no escape from paying the inevitable penalty to regain it.

"Practically, whether we like it or not, we cannot withdraw from Egypt nor abandon the Soudan. We are now there, and if we were to withdraw, some other first-rate Power would attempt to take our place; but the European Powers generally would never permit it, because England alone possesses interests sufficient to afford full security to all other Powers that an open route through Egypt and the Suez Canal shall always be securely maintained.

"Concurrently, however, with the clear enunciation of our intention to be the responsible custodians of Egypt, must appear to the people of that country our determination to preserve them from oppression, develop their resources, and employ them in all possible cases in the various administrations of government. Then, and not till then, we may safely proceed to send relief and other expeditions up the Nile

Valley, which I venture to predict will receive a welcome as they proceed, provided, of course, the people thoroughly understand that we are now their responsible protectors.

"JOHN FOWLER.

"2, QUEEN SQUARE PLACE, WESTMINSTER."

"*To the Editor of 'The Times,' May 21st, 1884.*

"SIR,—It is no longer possible to withstand the overwhelming and almost universal feeling in and out of Parliament that something effectual must be done, and without further delay, to rescue Gordon and England. The danger to be guarded against at the present moment is rather that in the impatient and excitable state of the public mind on the subject a mistake may be made in the scheme of rescue.

"It is indispensable, however, as a first step, that we should distinctly recognise the fact that no expedition for the relief of Gordon, however costly in men and money, could possibly reach Khartoum in time to accomplish its object, unless preceded by an authoritative proclamation of our intention to assume undivided and continuous responsibility and authority in Egypt.

"The generous offers of money and personal services you have received indicate the extent and depth of sympathy and shame which exist; but care must be taken that neither these volunteer nor official efforts are expended in a wrong direction, and it is in that view I venture to occupy once more a little of your space.

"One proposal which is recommended is for constructing a railway across the desert from Suakin to Berber. I venture to say that such a work would be a great mistake, because not only would it be exceedingly costly and difficult by reason of the general character of the desert and the mountain range which has to be surmounted, but it would leave completely unprotected and undeveloped the whole of the Nile Valley between Berber and Wady Halfa, which includes Dongola and Dabbe, the caravan entrances to Kordofan. Egypt has already had experience of railways constructed across waterless deserts in the line from Cairo to Suez, which was taken up and abandoned after struggling for some years against the ruinous loss of its working and its comparative inutility.

"It would be an advantage in every way, in my opinion, if Egypt was limited to the valley and delta of the Nile, and if the present opportunity was taken for the settlement of boundaries between Egypt and Abyssinia, and their strict observance secured effectually by the conditional transfer to Abyssinia of one or more much-coveted ports on the Red Sea. By this arrangement, and thus being able to confine operations in Egypt to the Nile Valley, the steps to be taken are simple, and attended with little or no risk. The value of the Nile for navigation greatly varies in different parts of the river, and at different periods of the year, and although rocky obstructions or cataracts are numerous in some parts, there are also long stretches of water which can be used advantageously as water carriage only.

"The experience I have gained in Egypt leads me to suggest and strongly recommend the immediate completion of a communication, partly by railway and partly by water, between Wady Halfa and Khartoum. From Cairo to Wady Halfa the existing route, partly railway and partly river, is fairly satisfactory, and may be further improved as regards the river by a short ship incline at Assouan and the removal of a few obstructions between Assouan and Wady Halfa, which I estimated, after careful investigation, at £3,000 a year for a few years. From Wady Halfa southward the more serious works commence, and from actual examination it was found that as far as Hannek (the Third Cataract), a distance of 200 miles, the river consisted of a succession of obstructions which render it useless for ordinary traffic. Over this ground, therefore, it would be necessary to lay down a railway, but the sections and surveys show that very little cutting or embankment would be required, and about 50 miles have already been completed. From Hannek the River Nile, by way of New Dongola, Handak, Old Dongola, and Dabbe to Ambukol, a distance of about 170 miles, can be again used as the means of communication, and although some rocky obstructions exist they may be gradually removed and the navigation improved as the traffic increases.

"Ambukol is known as the point where the Nile makes a vast bend to the east, round by Aboo Hammed (where the Korosko desert route joins the Nile), Berber, and Shendy to Khartoum, and not only does this bend of the river involve a great additional length as compared with a direct line, but it is so much interrupted by cataracts and rapids between Ambukol and Berber as to make it necessary to leave the river at Ambukol and make a railway direct to Khartoum through the Wady Mokatten, which is peculiarly favourable in every respect for the purpose.

"My proposition, then, would be to concentrate all rescue efforts for the present, both military and otherwise, to the improvement of the communication up the Nile to Khartoum, and with the view of saving time and money and being thoroughly prepared for vigorous proceedings in the early autumn, provide, without delay, special steamers and lighters for the river service, and rails, sleepers, locomotives, and other appliances for the railway works. The total length of railway to be made would be about 335 miles, and of river to be supplied with steamers and lighters about 200 miles, and the cost may be safely assumed to be within £2,000,000, including all materials, carriage, and labour, and the payment of native labour.

"Unlike all expeditions and expenditure purely military, the result of the expenditure and work I recommend would be available for all time as the means of encouraging trade and promoting civilisation, and during its construction it would largely employ natives who are much in need of the food they would receive.

"In September and October the steamers and all other materials may be ready, and as the Nile would be in flood, may ascend to their different destinations, and the works would then proceed rapidly.

"The peculiar advantage of Khartoum is sufficiently known, being at the junction of the Blue and White Niles, and in the centre of communication with Berber; indeed it would be impossible to exaggerate the value of a position which possesses such advantages. I feel this proposal is wanting in the dash and magnificence of other suggestions, but I know it to be practicable, and I believe it to be more economical, certain, and even more rapid than any other scheme.

"JOHN FOWLER.

"2, QUEEN SQUARE PLACE, WESTMINSTER."

In 1885 he was selected as Conservative candidate for Hallamshire, his native district of Yorkshire. On this occasion, as before, his chief line of attack on the Gladstonian party was an indictment of their foreign policy, more particularly their abandonment of the Soudan to anarchy and chaos.

The work of canvassing a large and populous division was soon found by him to be a great tax on his strength, and he very wisely, on the advice of his doctor, withdrew from the contest. The reader will agree with the congratulations sent to him by Mr. Dent, one of the directors of the Forth Bridge Railway Company.

"DEAR SIR JOHN," he writes, ". . . I am very glad that you have given up Hallamshire; the successful finish of the Forth Bridge is a greater crown to a man's life than a seat in Parliament."

His political campaigns, although never resulting in his return to Parliament, were in their way very successful. Although in private conversation he was wont to condemn his political adversaries with considerable vehemence, when he took the field as a candidate he rose to the occasion and put great control

on his habit of passionate utterance. The Tewkesbury
election of 1880 was conducted with complete good
temper. The defeated candidate was hardly less
popular than his successful rival. His carriage was
drawn to the station amid the cheers of both parties.
Though he did not secure their votes, he earned their
regard and respect.

All who knew Sir John will remember this very
human contradiction in his character. In his less
guarded moments he was apt to be impulsive,
passionate, and even arbitrary, full of the characteristics of the spoiled child of fortune. Such foibles,
when combined with a real kindliness of heart, merely
supply the light and shadow which endear men to
their friends. Fortunately for us all, friendship goes
by favour. There is probably nothing less provocative
of affection than that managed character in which all
spontaneity has been suppressed by a too rigid self-
discipline.

It was a peculiarity of Fowler's frank and vehement
nature that he could put himself under discipline when
he had some great objective in view. He was not a
man with whom friends, who happened to differ from
him, liked to talk politics. The warmth of his invec-
tive was apt to be contagious. When he went elec-
tioneering, however, he was respectful to opponents,
though their views, to his mind, were not a little disordered. In private he did not suffer fools gladly,
but he had the capacity for taking infinite trouble to
persuade them, when he met them on parliamentary
committees or boards of directors. He was most
careful to prepare his speeches beforehand, and when
important letters had to be written he would reject

draft after draft till he had got it quite right. He had, in fact, that great element of genius, a capacity for taking pains.

As a Highland landlord, he naturally took an interest in that most important political and economic problem, the future of the crofter population. In a speech addressed to the annual reunion of the natives of Ross and Cromarty, held, significantly enough, not in Ross or Cromarty, but in Glasgow, in December, 1885, Sir John touched on this vexed question :—

"Let me begin," he said, "by speaking of landlords as I would of babies, that as they are inevitable it is best to be kind to them. But I go further and say landlords are desirable, and especially in mountainous and thinly-populated countries, because it is well to have educated men and women in every district, and also because large and comprehensive works of improvement can only be carried out by their means. Landlords also have their friends who visit them and enjoy intercourse with the natives to the advantage of both. Of course it is essential that a landlord should be properly equipped for his duties in a knowledge of the work he has to do. At Braemore I have had the responsible duties of a landlord for twenty years, and during that time I have done much work and made many mistakes. I have, however, succeeded in improving all the arable land in the strath, so that it will now produce at least three times the food it was capable of producing when I first went to Braemore. . . . But a landlord must have tenants, and it so happens that I have at Braemore a large farmer, a small farmer, and crofters. My large farmer is my son. . . . My small farmer is Mr. Chisholm, who is the best road-maker in the north, and his wife has the largest family of small children, and feeds the biggest chickens of any woman in the Highlands. Their elder sons have wisely either emigrated or migrated. Two of them are, I believe, with us to-night, and are doing well in Glasgow. At Braemore I have also crofters, but fortunately for themselves and for me they all earn money outside their crofts, and the croft is valuable to give food for their cow and vegetables for the family. From a long experience of the Highlands, I am much disposed to think that many of the difficulties and differences which exist are due to want of accurate knowledge on the part of those who give opinions and interfere in these questions. I will give you two remarkable instances of this in my own experience. Some years ago a friend of mine,* a liberal, philanthropic man, and a member of Parliament, came to see me at Braemore. He had never been in the Highlands before, but he had been making speeches and writing articles on the millions of acres of land in the Highlands, which if cultivated would grow unlimited food for the people, but was now wasted on sheep, cattle, deer, and wild cats. Well, I took him over the property, and he saw for himself, and was satisfied that not one hundredth part of the ground was capable of the cultivation he had expected, and he found that every yard of ground which was cultivable was cultivated. He was astonished, but he was honest, good, and true, and he thanked me again and again for saving him from repeating his mistakes; and yet to this day you hear constantly from ignorant or mischievous persons the same gross misstatements. Another mischievous misstatement is with respect to deer forests, as if all the land of the high mountains of the Highlands would produce mutton or beef. Now it so happens that adjoining a part of my property at Braemore is the property of Inverlael, which has long been in the hands of my excellent neighbour, Mr. Walter Mundell, as a sheep farm. A great portion of Inverlael is more than 2,000 feet above the level of the sea, and some parts more than 3,000 feet. Probably a more competent and experienced farmer does not exist than Mr. Walter Mundell, and yet the high death-rate of his sheep on that elevated ground, combined with a long continuance

* The late Mr. Samuel Morley.

of low prices, rendered it quite impossible for him to continue it as a sheep farm without destructive loss, and of course such ground was far too high for cattle. Now as a practical common-sense question, what is the proper course to adopt in such a case? Is the land to be abandoned and remain useless to the landlord, and afford no employment to the people, or to be let as a deer forest, and bring employment and other advantages with it? Surely there can be only one answer to such a question. Well, ladies and gentlemen, we have seen how landlords, sheep farmers, and crofters may be comfortable in Ross-shire, and of course in other places also, and we have seen that Highlanders may prosper in our colonies, and we know that they prosper in Glasgow and London and all the great cities of Great Britain, and there is ample room for all the surplus population. Now, don't you think that under these circumstances that wise heads and kind hearts ought to be found to settle all differences which now unfortunately exist in the Highlands?"

A practical summing up of the question which seems to us to leave little more to be said.

As a master and employer of labour Sir John Fowler was very popular. Many proofs of this have been forthcoming in a quite spontaneous manner. "I was over thirty years in his service," said one trusted servant at Braemore, "and I do not think I ever had a cross word from him. Indeed, he was more like a father to me than a master." The same informant dwelt with warm appreciation on the interesting talks he had had with his master in their long walks in the forest. No one is more capable of valuing such glimpses of the great world than the intelligent Highlander, whom fate has retained in his northern home. Another, a coachman, long since gone into other employment, has narrated how when it was his duty to drive Sir John

to the distant Garve Station, starting at three or four
o'clock on a late autumn morning, his master would
personally see that the servant as well as himself had
a good breakfast before starting, and all in a manner
so kindly and thoughtful that the remembrance of it
was cherished after long years. In many of his
letters to Lady Fowler he expresses his gratitude for
the attention and care (not, so far as we can gather,
of any extraordinary nature) of the servant who
happened to accompany him. So, too, with his own
associates he seemed to look on what many regard
as the mere ordinary courtesies of life as real kind-
nesses to himself. This somewhat eager recognition
of personal attention was a genuine source of content-
ment, a trait of character calculated to make a man
very popular, even though at times it may have
made him tolerant of flattery. Full of goodwill
himself, he looked for and appreciated the goodwill
of others. Yet on occasion he could be exacting.
If a thing was wanted, neither trouble nor expense
was spared to get it at once, and of the best quality.
Still he was a man whom it was a pleasure to serve,
and with whom to exchange courtesies was a grateful
office.

Apart from the sports and duties of his Highland
home, Sir John's recreations and interests are evidence
of a liberal taste. They were active and social rather
than sedentary. He collected, much aided therein
by Lady Fowler, a valuable collection of pictures.
After his death his famous Hobbema fetched a
record price at Christie's. He had a number of
Turner's water-colours, and his appreciation of fine
work was real and discriminative.

He was not a great reader, though being a bad sleeper he often had some volume of light literature by his bedside at night. He kept himself informed of the advance of the sciences connected with his profession, partly by reading, but more often by conference and talk in the daily round of his work.

In connection with his Highland home, where he naturally had more leisure, he read with great interest topographical and antiquarian works. Thus, one of our informants remembers to have found him busy with Sir Thomas Dick Lauder's *Account of the Great Floods of August, 1829, in the Province of Moray and the Adjoining Districts* (Edinburgh, 1830), a work full of curious information for a Highland laird who happened also to be an engineer. Burt's *Letters from the North* were also in his hands at this time. Captain Burt was an engineer officer who, about the year 1730, was sent to the north of Scotland as a contractor; he gives in a series of letters a most curious account of the poverty of the Highlanders of that period, as well as of the arbitrary power of their chieftains. Sir John also had by him for constant reference a set of the *Annual Register*, a storehouse of facts from which his love of detail and comparison extracted a vast amount of pleasure and amusement. This statement is sufficient to show that his recourse to books was rather that of the lover of action and outdoor life than of the student of literature.

We noticed on an earlier page his diffident dislike of speaking in public, and in his earlier days no amount of preparation seems to have made him quite comfortable on a public platform, a somewhat strange contradiction in one naturally so self-confident. In later years this distaste wore off, and the specimens already given show that he spoke with enjoyment and humour. Indeed, participation in public meetings became in his later life almost a favourite recreation.

To several well-known institutions and societies his practical common sense and tact proved of great service. To serve as chairman to a body of Egyptologists might seem to the uninitiated something of a sinecure. We are assured, however, that breezes can blow even in the placid back waters of Egyptology, and Sir John's moderating influence, we can well believe, was invaluable in getting business done. The following communication from Sir E. Maunde Thompson, of the British Museum, is an estimate from a competent hand of Sir John's serviceableness in this respect.

"My acquaintance with Sir John Fowler began with his election as President of the Egypt Exploration Fund at the close of the year 1887, an office which he held down to the time of his death. The choice of Sir John Fowler to guide the destinies of this rising society was a happy one. He was not an Egyptologist, he made no pretence of any special knowledge of the history and archæology of ancient Egypt, and it might have been argued that a president of an archæological society should himself be an archæologist. But the learned man does not always combine with his learning those practical qualities which are so essential in administration ; and in a young society, such as the Egypt Exploration Fund then was, business capacity in its leaders is of paramount importance for its ultimate success.

"Still, if he was not an Egyptologist, Sir John Fowler was interested in no small degree in the progress of things in Egypt, and his engineering connection with that country and his practical ability were reasons for selecting him to be president of a society formed with the view of exploring the antiquities and ancient sites of that mysterious land. His success in piloting the fortunes of the Egypt Exploration Fund for eleven years, during which it rose to be one of the most prosperous and most energetic of the archæological societies of Great Britain, was a source of satisfaction to others besides himself. He was pre-eminently a man of that 'saving common sense' which smooths away difficulties and reconciles conflicting opinions and guides affairs into the path which leads to prosperity.

"His amiable character contributed in no small degree to the great success of the society's undertakings. For who can get the better of one who is always good-tempered, always fair, always courteous, and, moreover, always firm?

"During Sir John Fowler's presidency the Egypt Exploration Fund carried out a series of undertakings which may be equalled, but can hardly be surpassed, in the future. In the years 1887-9 the sites of Bubastis and the city of Onias were explored; and next followed excavations at Ahnas ro Heracleopolis Magna in 1889-91, both yielding important sculptures and antiquities, now to be seen in the galleries of the British Museum and of other museums, both in this country and in America. But the greatest work of the fund that has signalized the period of his presidency is the clearance and restoration of the great Theban temple of Dêr el-Bahri, near Luxor—a herculean labour which has taxed the energies and the resources of the society since the year 1892. And while this great work has been in progress the excavation of minor sites has not been neglected, and further, a valuable series of publications descriptive of the discoveries has been issued. Again, in the years 1888-9 the society sought a further outlet for its energies in the organisation of a branch known as the Archæological Survey, which, among other results, has published an excellent series of volumes on the famous tombs at Beni-Hasan. And, lastly, the discovery, in the course of excavations in the Fayûm in 1896-7, of an enormous mass of papyri of the Ptolemaic and Roman periods, which was unearthed at Oxyrhynchus, has resulted in a

2 A

further extension of the field of labour and the foundation of a Græco-Roman Branch of this flourishing society.

"I know that Sir John Fowler regarded his connection with the Egypt Exploration Fund, and the prosperity to which it attained under his direction, with legitimate pride. And indeed he had full reason to be proud. I trust that it may not be thought presumptuous in me, who saw so much of him in connection with the administration of this society, and who thus learned to admire his simple and manly character, to have written this imperfect appreciation of the debt of gratitude that all who take an interest in the art and archæology of ancient Egypt owe to Sir John Fowler."

It is not given to many to pass beyond the limit of threescore years and ten, without feeling the heavy and disabling hand of time. Sir John, although a hard worker, was also careful of his health, and was happily allowed to enjoy a green and vigorous old age. The following letter of reminiscence which has been kindly sent to the author by Sir Benjamin Baker is a valuable contribution to the philosophy of wise husbandry in the matter of health, as well as an interesting addition to the biography of his friend and partner :—

"You ask me," he writes, "if any of Sir John Fowler's early professional friends are likely to be able to contribute information respecting his career. Sir John outlived most of his professional friends. Stephenson, Locke, Brunel, Rendel, and other leaders all died before they reached the age of fifty-seven. Sir John lived thirty-three years after occupying the Presidential Chair of the Institution of Civil Engineers, whilst on the average other presidents have so far survived no more than thirteen years after attaining the same distinction in the profession. The early death of so many leading engineers was accepted by their colleagues as a warning that the anxiety and responsibility inseparable from a great engineer's work cannot be safely associated with continuous application to their duties and no holidays. At an unusually young age, therefore, Sir John made it an annual duty to take a long holiday in the Highlands, where his thoughts were diverted into an entirely different channel, and this practice finally led to the acquirement of the forty square miles of hills, straths, rivers, and lochs in Western Ross, and the transformation of the whole into the beautiful estate which is well known throughout the North as Braemore. The same regard to his health led him to place himself in the hands of doctors as regards diet and other matters to an extent which amused his friends, knowing as they did with what a robust constitution he had been endowed. He might easily excite the sympathy of strangers by being seen in Westminster with a woollen scarf over his mouth as a respirator, but his friends would know that perhaps a few days later he might be found *minus* respirator, lying amidst sleet and rain on the bleak top of one of his own high hills at Braemore, waiting patiently for hours for the hinds to move on, or for a stag to rise, and returning in the dark after a long day's stalk without any thought of hardship, provided he brought a fine beast home with him, which he generally did.

"My first visit to the Institution of Civil Engineers was to hear my late partner deliver his presidential address in 1865; and his last visit was to perform the duty devolving upon him as the oldest surviving Past-President of proposing a vote of thanks to me for the presidential address delivered on my taking the chair occupied by him thirty years before. The growth of the engineering profession during the aforesaid interval is well illustrated by the increase in the number of members of the Institution from 1203, on the occasion of the first address, to 6,907 on that of the second.

" I first met Sir John Fowler forty years ago, as a youngster about to proceed to India for a career. The impression I then formed of him is as clear to-day as ever, and has never varied. I felt myself in the presence of a born commander of men, who formed his opinion with instinctive quickness, held on to it firmly, never questioning its soundness himself, nor failing sooner or later to satisfy most of his hearers by ingenuity of argument and charm of manner, that if they held a contrary opinion originally it was a fortunate thing for them that they had come to see him before it was too late. He advised me not to go to India. I thought he was wrong, but obeyed. I have long since known that he was right.

"In many far more important matters his great experience and rapid and comprehensive grasp of all of the circumstances have, to my knowledge, led him to give advice, the rejection of which has afterwards been seriously regretted. Thus the Great Western Railway directors long remained undecided on the question of bridge *versus* tunnel across the Severn. Sir John strongly advocated a bridge, and we prepared designs and estimates, which were thought too bold and risky at the time, but which our subsequent experience at the far more difficult work across the Forth more than justified. He told the directors that a tunnel would take twice as long to construct, would cost far more than the estimate, on account of trouble with influx of water, and for the same reason would cost more to maintain, as the tunnel would require to be continuously pumped, whilst the bridge would want nothing but a coat of paint. Most important of all, however, the bridge would carry safely far more traffic than a tunnel by reason of the better gradients and facilities for intermediate block stations. All this has proved to be true, and there is little doubt that the present directors regret that their predecessors rejected the advice of their consulting engineer, and handed over to them a tunnel instead of a bridge. Similarly, to my knowledge, the Great Western Railway would have been at Southampton Docks, and the Great Northern Railway would years ago have effected amalgamations which would have obviated the difficulties arising from the costly extension of the Manchester, Sheffield and Lincolnshire, or Great Central Railway to London, had they followed Sir John's advice. It is hardly necessary to say that qualities such as Sir John Fowler possessed would have ensured success in any profession or business, and it is a question whether his country would not have benefited more by his services as a great organiser and leader of men in warlike operations than in the peaceful pursuit of railway making.   Leaders of men necessarily invite criticism; their personal characteristics often present many different phases, and the opinions of casual critics may reasonably vary according to the particular phase of which they have had experience.   At one time of his life Sir John was much troubled with distressing headaches, and being at the same time hard pressed with work, I can readily understand that anyone, knowing him only when the headache was in force, would form the wrong opinion that he was irritable and unreasonable in manner and judgment.   All who knew him well, including his engineering staff, knew that these stormy intervals were but as a few claps of thunder in a summer's season, and that his normal attitude was one of infectious cheerfulness, for which everyone having to do with him felt the better.   Occasionally, however, somebody would resent a remark or action of his, and a slight initial difference between two strong and proud men often widens into a serious breach of former friendly relationships.   Such breaches, of course, did occur in Sir John's long career, as they have in that of all other prominent men's, and they have sometimes led temporarily to wrong inferences, which have long since proved themselves to be unfounded.

"For many years past no secrets, personal or otherwise, existed between my partner and myself, as we consulted each other on even the most private personal matters.   I claim, therefore, to know his character 'down to bed-rock.'   His last letter to me, written that I might be 'the first to know' that he was sick unto death, was subscribed 'Your affectionate friend,' and I am proud to have held and to have returned for more than thirty years the friendship and affection of such a man."

For many years Sir John Fowler had suffered from attacks of bronchitis, the result of chills caught on some of the frequent journeys necessitated by his

profession.   Gradually this caused weakness and dilata-
tion of the heart, and, during the last few years of his
life, an increasing difficulty in breathing.

Notwithstanding his growing infirmities, he continued
to take a close interest in much of the work which went
through the office at Queen Square Place.   The last
matter of importance to which he gave personal supervision was the difficult and costly engineering task of
crossing the river Findhorn, on the new section of the
Highland Railway between Aviemore and Inverness.
It had been originally proposed to take a circuitous
route, and to build near Tomatin a bridge of very
moderate dimensions.   Sir John, however, succeeded in
persuading the directors to adopt a more direct course,
and to save thereby a distance of between one and two
miles.   The present lofty and magnificent viaduct is
the result.   The stone used in the fine masonry of this
work came from the quarries which had supplied granite
for the Forth Bridge, and in excellence of this kind the
veteran engineer found an intense delight.

In July, 1896, when he had attained the ripe age
of seventy-nine, he became for the first time seriously
ill.   In August, however, he was able to go to his
beloved Braemore, and, as usual, to entertain his
friends.   He returned to London in October.   There
he remained till, in January, 1897, by his doctor's
advice, he went to Egypt, accompanied by Lady
Fowler and a medical attendant.   The air of the
desert and a quiet rest at the Mena Hotel, near the
Pyramids, revived him, and on his way home he was
able to enjoy a six weeks' visit to Hyères, where he
found his old friend and partner, Mr. Baldry.   After
reaching home he had a relapse in June, and, though

THE FINDHORN VIADUCT.

To face page 358.

able to go to Braemore, it was evident to himself and to those who loved and watched him that he was growing weaker. It was at this time he wrote the touching letter to which allusion is made in the above - printed communication from Sir Benjamin Baker. One of the last visitors at his Highland home, Mr. Alfred Gathorne-Hardy, remembers well how shocked he felt at the changed appearance of his kind friend and host. Sir John, however, would not hear of the visit being curtailed, and though obliged to keep his room on many days he followed with kindly interest the sport of his guests in the forest and on the river.

In October, 1897, he returned to London, where he became very ill. His medical adviser, Dr. Kidd, for whose sedulous care he frequently expressed the warmest gratitude, became afraid in February, 1898, that the end had come. The family was summoned. His strong spirit, however, once more bore up against disease, and he rallied so far that he was able to go to Brighton for a fortnight in May. His sister, Mrs. Whitton, now a widow, at this time returned from Australia. To her, and indeed to all the members of his family, he had ever been warmly attached, and the meeting was a source of much quiet joy to the brother and sister. He continued fairly well, though extremely feeble, during the summer. In September he went to Bournemouth, where, until the end of the month, he was able to go out daily in an invalid chair. He grew, however, visibly weaker, but at intervals he recovered his clear mind and fortitude, and talked with calm resignation of that which was to come.

On November 20th, 1898, he was feeling unusually bright, and joined his wife and his brother, Mr. Frederick Fowler, at dinner. He asked his brother to tell other members of the family that he was in full possession of his faculties, and presently he retired to his room, which adjoined the dining-room. Here, seated on a sofa, while his favourite servant, Frederick Theyer, was helping him to undress, he suddenly and very peacefully passed away. His work was done, and, full of years and of honour, he was vouchsafed the crowning mercy of a painless death.

The work of men who have excelled in the arts of peace does not touch the imagination as do records of conquest and stirring adventure, but it has a permanence and importance which is not to be gainsaid.

The life-work performed by Fowler and his compeers, by the relatively prosaic methods here described, has been revolutionary in its consequences. The England through which Fowler travelled to London in the "Emerald" coach in the year 1838 is not the England which he left in 1898. Yet, so numerous have been the leaders of the army marching forward in step, so general, so simple, and so obvious has been each advance, that we find it difficult to perceive and to insist on the personal element in this vast change.

The effacement of the record of great industrial captains will, on reflection, enhance rather than detract from their title to the gratitude of mankind. It is true that, in the life of Stephenson, as depicted by Dr. Smiles, the figure of the hero emerges out of the chaos in which lurked his great opportunity, homely, resolute, and successful in its appeal to our historical imagination. He seems to pass, and to carry the world with him, at one step out of the darkness of the middle ages into the modern economy of an industrial state. The biographer is well within his rights when he dwells even to exaggeration on the width of that stride. Yet even Stephenson, who has been identified with the locomotive engine, had his precursors. His fame, moreover, is great and lasting, not so much because of his own work, but because of the vastly superior work which his successors have been able to engraft on his achievement. All these considerations have a levelling effect. It would almost seem that those who work with, rather than against, the stream of progress, are, by the very success of their effort, doomed to a certain supersession and oblivion. After Stephenson, a typical figure to whom much is assigned that happened both before and after his time, many opportunities for improvement of transport have been seized, but to none of them has the public vouchsafed the same high acknowledgment of fame. After Stephenson it was, for a time at any rate, to be the day of the great organiser rather than of the great inventor; and it is as a great organiser that Fowler becomes a pre-eminent figure in the industrial revolution of the century.

These levelling reflections detract in no way from the merit of those who receive and utilise, who rekindle with increased glow, and then hand on the torch of knowledge and enlightenment from generation to generation. On the contrary, they enable us to discriminate, to appreciate the value of services which have been eclipsed by the very greatness of their results, and to set them in contrast with those barren heroisms which leave their perpetrators in splendid isolation. The great man, in Carlyle's view, is too often the man of violent and unscrupulous character, who is borne into a place and power on some backward eddy from the otherwise even current of human progress, by an appeal to the ignoble instincts of fear, superstition, and force. The Mohammedan power, the crushing burden of militarism under which Europe still groans, are gifts which we owe to Mahomed, and to such as Frederick and Napoleon—men whose abnormal opportunity and personality Carlyle has characterised as heroic.

Happily greatness of this kind is as rare as it is monstrous. It can have few imitators, but remains in isolation, a solitary monument to dazzle the imagination of succeeding generations.

The greatness achieved in the arts of peace is in every way a contrast to that of the Carlylean hero. Under free institutions the coercive hero has no permanent usefulness. The able man, the great man, the worker of beneficent revolution, is he who opposes superstition, and vicious but inveterate custom, by some new and convincing exposition of right reason; who, like Bentham, from his philosopher's study, a figure not ordinarily regarded as heroic, exercises a silent but irresistible influence on the laws and administration of his country; who, like Adam Smith, gains the assent of his countrymen, or even of a larger audience, to some just law of social development; who, like Newton or Stephenson, discovers some hitherto neglected law of nature, and applies it for the enlightenment or service of men; or, to come to the subject of the present biography, who takes a yeoman's share in domesticating for the use of man the new discoveries of science, the triumphs of hydraulic, steam, and electrical engineering. These, and such as these, are the great qualities of civilisation. They make no appeal to the fratricidal impulses of force and fraud which lie buried, not very deeply, in human nature, and which seem to give to him who evokes them on a large scale the isolation of greatness. They rely rather on the equity of free contract, and on the self-discipline and independence of labour which is thereby engendered. They recognise that, in accordance with the great law of the economy of effort, slavery, arbitrary power, and the oppression of labour are things morally and economically inconsistent with modern conditions.

Men possessed of, and called on to exercise, these qualities are not isolated from or placed above our industrial system. They are part and parcel of it. They are the moving spirits in an economic harmony, wherein the ceaseless energy of human effort is spontaneously guided into honourable and serviceable channels.

We are far, alas! from an ideal condition of industrial life, but who can doubt that progress must be organised by the expansion and adaptation of this principle!

It is in the light of these considerations that we claim for men of scientific and industrial achievement a meed of honour more real and more due than that which is given to great conquerors. England's greatness among the nations—the greatness, that is, of modern England—will never in the eye of history consist in mere supereminence of material empire, but in her pioneer advance, scatheless, through the

APPENDIX

ADDRESS OF JOHN FOWLER, Esq.

President of the Institution of Civil Engineers,

ON TAKING THE CHAIR, FOR THE FIRST TIME AFTER HIS ELECTION,
JANUARY 9, 1866.

GENTLEMEN,—On assuming the chair of this Institution as its President, and undertaking for the first time its duties and responsibilities, allow me to assure you that I feel deeply the honour you have conferred upon me by electing me to this, the highest position to which the Civil Engineer can aspire ; and that I feel still more deeply the weight of the duties which are inseparable from this honour. I will also venture earnestly to request you to extend to me your indulgence during my period of office, and afford me your kind co-operation in any efforts I may make for the advancement of our profession, or for increasing the usefulness of this Institution.

I ask this assistance from you with peculiar anxiety, because I cannot but feel that the present is a period of unusual importance to this Society, and that the rapidly increasing prominence of the profession demands at our hands a corresponding care for its efficiency and dignity.

The high degree of material prosperity which this country and its dependencies have now happily enjoyed for a considerable time, has naturally led to great activity in our profession ; probably at no former period have the skill and enterprise of engineers been so severely taxed as during the last few years ; and as civilisation continues to advance, and labour to require increased assistance from mechanical contrivances, the connection of civil engineering with social progress will become more and more intimate.

I hope I may be allowed to say, however, with a deep feeling of professional pride, that hitherto the inventive genius, the patient perseverance, and indomitable energy of the members of our pro-

fession have not been found unequal to the tasks they have been
called upon to perform; and although I have full confidence in the
future, I venture to suggest that the present is a fitting moment for
considering the means by which our younger brethren may be best
prepared for the arduous duties, and growing difficulties, which they
will undoubtedly have to encounter in their professional career.

It is not merely that works of magnitude and novelty are increasing, and will continue to increase, but it is becoming apparent
that we shall have to meet the competition of foreign
engineers in many parts of the world; and that great
efforts are now being made, not only by careful scholastic
education, but by more attention to practice on works, to render
the civil engineers of France, Germany, and America, formidable
rivals to the engineers of this country.

Competition
by foreign
engineers.

Here it has always been found that friendly and honourable
rivalry among members of the profession has been on the whole
beneficial to science and to engineering progress, and we cannot
doubt that the same result will follow the more extended rivalry
which we shall have now to meet from the engineers of every
nation. At the same time this consideration renders it our especial
duty to take care that the distinguished and leading position which
has been so well maintained by our great predecessors, shall not be
lowered by those who come after them.

My predecessors in this chair have addressed you chiefly upon the
interesting topics and works of their own time, and with so large
a field demanding their attention, it was natural that
they should devote themselves mainly to describe the
past, and to indicate in outline the features of greatest interest in
the present.

Former
addresses.

My immediate predecessor, Mr. McClean, gave to the Institution
a description of the remarkable results which had been produced by
the general introduction of railways into England in combination
with steam power, and clearly pointed out their influence on the
increase of its material prosperity and national wealth.

Mr. Hawkshaw pointed out the rapidly increasing importance of
wrought-iron for engineering works, with the promise of new applications of steel; and the fact and consequences of the increasing
speed of railways and steam boats.

Mr. Bidder, after defining the object and scope of the profession
of the civil engineer to be "to take up the results discovered by the
abstract mathematician, the chemist, and the geologist, and to apply

them practically for the commercial advantage of the world at large," illustrated his views by selecting the examples of hydrodynamical science and hydraulic engineering, for the purpose of pointing out the serious mistakes which might result from the neglect of a proper knowledge of true mathematical principles.

Mr. Robert Stephenson described the modern railway system in England up to the period of his address, commenting upon its extent, and justly appreciating its value; and he reviewed, in a large and philosophical spirit, its system of management, and the commercial economy which it had produced.

Mr. Locke in like manner selected for his subject a description of the French railway system and its management, in the introduction of which he had himself been so actively engaged.

Another of my predecessors, Sir John Rennie, seems to have been determined that no single topic of professional interest should remain to any future President which he had not himself exhaustively discussed; for he not only presented a complete panorama of all past engineering works, but he gave a descriptive analysis, so full and complete as to make his address at once a history of engineers, and a manual of engineering science.

The whole field of discussion and description of the past has thus been so completely and so ably occupied by my predecessors in this chair, that I shall not attempt to travel over the same ground; but I propose to deal almost exclusively with the future, and endeavour, although I possess no peculiar personal fitness for the task, to suggest some of the means by which the younger members and the rising generation may best prepare themselves for the duties which that future will bring with it.

The Future of the Profession.

I may first briefly notice, and for the purpose of illustration and introduction, a few of the great engineering problems of remarkable boldness and novelty which are now presenting themselves for the supply of the future wants and convenience of mankind: amongst them may be enumerated the Suez Canal; the tunnel through, and the railway over, Mont Cenis; railway bridges over and under great rivers and estuaries; new ferry works of unusual magnitude; vast warehouses and river approaches for commercial cities like Liverpool; railways under, over, and through great cities; long lines of land and ocean telegraphs; and comprehensive schemes of water supply, drainage, and sewerage.

The Engineering Problems of The Present Time.

All these works present problems of great interest; and it will require cultivated intelligence, patient investigation, and enlarged experience, to accomplish the task of their satisfactory solution.

For the Suez Canal we must be content to wait a few years before the work be so far advanced as to enable us to judge of the effects of the physical and moral obstacles which to some experienced minds have appeared all but insuperable.

The Mont Cenis Tunnel, and the temporary railway being constructed over its summit, will continue to be watched with interest by all engineers, and it may yet be a question how far the mode of traction which has been adopted for the temporary railway will prove to be the best. The modified locomotive with the aid of a central rail has no doubt succeeded in surmounting gradients which have hitherto been considered to be more severe than compatible with the economical use of the locomotive engine; but further experience is still required, and the results of the trial will be watched with great interest, because it cannot be doubted that conditions will continue to present themselves to which the ordinary locomotive engine cannot conveniently be applied.

In many of the proposed and future designs of bridges over or under great rivers and estuaries, no novelty in the principles of construction may probably be required, but in other cases the mere magnitude alone will demand new arrangements and combinations; and may possibly also suggest the use of steel for parts or the whole of the structure.

The docks and warehouses of our great commercial cities are rapidly advancing in importance, and are constantly demanding increased facilities to enable them to meet the exigencies of trade; and for this purpose every possible resource of steam machinery, and hydraulic and pneumatic mechanism, will have to be taxed, to obtain convenient and adequate power and expedition.

The new scheme of river approaches at Liverpool is one of the most remarkable proposals of modern times for its boldness in grappling with the difficulties and necessities of a pressing want, and the complete solution of a difficult problem. It is understood that the engineer of the Mersey Board, who has designed this great work, is preparing a model on a large scale, which I have no doubt will be brought before the Institution.

The railways under, over, and through great cities are amongst the most striking results engendered by the necessities of rapidly increasing and closely crowded population, and may be regarded as one of the most useful economical developments which engineering has supplied to satisfy the requirements of modern civilisation. The engineering problems they present are infinite in their number, and interestingly intricate in their character.

Ocean telegraphy is yet in its infancy, but enough has been done by the numerous lines already laid, and by demonstration before this Institution, to prove that further experience alone is wanting to enable deep or shallow sea cables to be successfully laid and maintained wherever they may be required; and probably in no branch of our profession is the future of greater interest than in the coming tele-graphic connection of every part of the world by sea and land, and in the political, commercial, and social results which must follow such a remarkable increase in the facility of general intercommunication.

The rapid growth of communities, to which I have already alluded, has also developed the necessity of provision being made for a more abundant supply of pure water, and for a more complete system of sewerage than is now generally possessed by our towns and cities. Some of these works are already being carried out, or seriously contemplated, on a scale of almost startling, but not unnecessary, magnitude.

It is plain, therefore, that in every department of civil engineering the wants of commerce and society are pressing more and more urgently upon the resources of our profession. We have ship canals, but the Suez Canal throws them all in the shade. We have long tunnels through our English mountains, but we have now to penetrate the Alps. We have large bridges, but larger are required. We have noble ports, but they are choked with trade, and new accommo-dation of an improved kind is called for. We have steam ferries across rivers, estuaries, and straits, and rapid ocean steamers, but higher speed and better accommodation are demanded. We have large warehouses with convenient mechanical appliances, but larger warehouses and better mechanical appliances have become a necessity. We have many thousands of miles of telegraphic communication, but nothing short of its universal extension will suffice.

In the solution of these problems, thus rapidly indicated, and in others which could be easily adduced, we may rest perfectly satisfied that the difficulties they present are not to be overcome by a stroke of genius or by a sudden happy thought, but they must be worked out patiently by the combination of true engineering principles, ripe experience, and sound judgment.

Having thus called your attention to the peculiar position of our

2 B

profession in consequence of its rapid growth, and pointed out some
DEFINITION of the problems which await an early solution, I shall
OF A CIVIL now attempt to describe the nature of the functions
ENGINEER. of the modern CIVIL ENGINEER; and consider how the
coming generation can be best prepared for its inevitable work, and
to what extent this Institution can be made ancillary to that purpose.

Although we know from history that men have existed from the
earliest times who have been distinguished by great mechanical
capacity, remarkable skill in working materials, profound science,
and constructive knowledge, yet it is only during the present century
that civil engineering can be considered to have become a distinct
and recognised profession. Now, however, it has assumed the
position of an *art* of the highest order. Perhaps we may without
arrogance be entitled to claim for it the title of a true *science*.

Many attempts have been made to define and describe a civil
engineer in a few general words, but all such attempts have been
more or less unsatisfactory. Still, though it is difficult, if not impossible, to describe an engineer by a short definition, it is not so
difficult to enumerate and describe the nature of the works he is
required to design and execute, and the professional duties he is
called upon to perform.

He has to design and prepare drawings, specifications, and
CLASSIFICA- estimates, and to superintend the carrying out of works
TION OF THE which may be thus enumerated :—
WORKS EN-
TRUSTED TO 1. Railways, roads, canals, rivers, and all modes of
A CIVIL inland communication.
ENGINEER. 2. Water supply, gas-works, sewerage, and all other
works relating to the health and convenience of towns and cities.

3. The reclamation, drainage, and irrigation of large tracts of
country.

4. Harbours of refuge and of commerce, docks, piers, and other
branches of hydraulic engineering.

5. Works connected with large mines, quarries, ironworks, and
other branches of mineral engineering.

6. Works on a large scale connected with steam-engines, with
machinery, shipbuilding, and mechanical engineering.

This list, which might be almost indefinitely extended, involves
a vast variety of work, and must appear almost appalling to a young
engineer. Yet it greatly concerns his future success that he should
as far as possible be prepared to undertake any or all of the works
embraced in the list.

I believe the personal history of most of us would show that circumstances have led us in a widely different direction, in the exercise of our profession, from that which we originally contemplated, and that the success of many men may be distinctly traced to their ability to avail themselves of unforeseen opportunities to advance in some new direction.

The civil engineer must therefore be prepared for the various classes of constructive works thus enumerated; but in addition to this professional preparation, it is of the first importance, Study of as affecting his true position, and the confidence which objects, and ought to be reposed in him, that he should also have their value. a correct appreciation of the objects of each work contemplated, as well as their true value, so that sound advice may be given as to the best means of attaining them; and he must be prepared, if necessary, to advise his employers that the objects which are sought are not commercially worth the cost of the means which would secure them. It is not the business of an engineer to build a fine bridge or to construct a magnificent engineering work for the purpose of displaying his professional attainments, but, whatever the temptation may be, his duty is to accomplish the end and aim of his employers by such works and such means as are, on the whole, the best and most economically adapted for the purpose.

The first question which will present itself to an engineer with respect to any proposed work, will be the selection of his material; and as this question is so vital to the accomplishment of Choice of a satisfactory result, I propose to treat it in a preliminary materials. and special manner. I wish to impress upon every young engineer a due sense of its importance, because probably a greater number of mistakes have been made by the use of a wrong material, than from any other cause.

In the case of stone work, it is essential that the mode of construction shall have reference to the character of the stone; and this requires much of the knowledge of the geologist, the Stone. stonemason, and the quarryman, so that the engineer may know how best to work and set the stone, and what are the peculiarities of the quarry as to its sound or unsound beds; and, in addition, he should have sufficient chemical knowledge to detect any unfitness in the conditions of use to which he proposes to subject the stone.

Let us always bear in mind, in connection with this subject, the example of Sir Christopher Wren, engineer as well as architect, who himself selected the quarries, and sometimes even the blocks of which his structures were composed.

Of bricks I must be content with saying, that the power of detecting the good from the bad, the suitable from the unsuitable, must be acquired by the combined assistance of reading, experiment, and practice.

Brick.

A knowledge of lime, and the art of making the best practicable mortar from each description of building lime, is almost of equal importance to that required for selecting the stone, brick, and building materials themselves, but it is somewhat remarkable that the art of preparing mortar in a proper manner is not so general as it deserves to be; and to secure good mortar is a matter of continual anxiety to the engineer.

Mortar and cement.

Mortar for engineering works is ordinarily made from cement (chiefly Portland cement); or from hydraulic lime, such as lias; or from ordinary lime, such as grey or chalk lime.

Cement is chiefly used in combination with sand in various proportions, according to the nature of the work to be executed, and it is not only necessary to possess the requisite knowledge and experience for determining the proper proportions of cement and sand for each individual case, but it is desirable to have the means of determining by direct and repeated experiment the strength and quality of the cement which it is intended to use.

In the case of hydraulic lime, such as lias, the same general knowledge of the proper proportion of sand to be used is also requisite, but, from the great variation in the character of lias lime, and the different proportions of silica and alumina in combination with the lime itself, it is essential to obtain a careful chemical analysis, in order to avoid the great disappointment and bad consequences which may result from ignorance of the various qualities.

Of ordinary limes, it is only necessary here to say that they are of almost infinite variety as to quality and constituent parts, and must each be dealt with accordingly; and the engineer can scarcely take too much trouble to inform himself of the exact nature of each lime he has to use, and the best mode of using it.

Modern science, and the convenient manner in which steam-power can now be applied, have given to the modern engineer the means of obtaining better mortar from the same materials than was possible before the general introduction of steam.

The heavy rollers and iron pan worked by steam-power are now almost universally used for grinding and mixing lime and sand for works of magnitude : they produce with properly proportioned ingredients a mortar so good in quality, and so equal in the time and power of setting, that the engineer can calculate with certainty upon the conditions under which his designs will be carried out; and when he has become thoroughly acquainted with the quality and power of good mortar, and acquired confidence in its use, he will feel himself justified in its adoption in cases where our predecessors, and even some modern engineers, would have hardly ventured to employ it in the place of the more costly Portland cement.

When iron is intended to be used in structures, it is essential to know under what circumstances cast-iron is best for the purpose, or when wrought-iron should be employed, and also when steel must be resorted to. The profession has probably been assisted to a greater extent by the experiments and writings of its members and of distinguished men of science in the material of iron than on any other subject; but these valuable investigations and experiments must be supplemented by the practical knowledge which can only be acquired by attentively studying the peculiarities of material and manufacture.

Iron.

Cast-iron or pig-iron remelted and run into moulds is largely used by engineers for columns and other parts requiring great power of resisting compressive strains ; and, as its price per ton is generally about one half of that of wrought-iron, it becomes a matter of economic importance to adopt it in all cases where it can be safely and properly used, but it is of the most varied quality and strength, and the greatest attention of the engineer is required to secure the proper kind.

Wrought-iron is perhaps less varied in its quality than cast-iron, and for many purposes of engineering it is the safer metal to adopt, from its greater power of resisting tensile strains, and less liability to sudden fractures. But it must be remembered that wrought-iron is sometimes pure and of high quality, sometimes very impure and of the commonest quality, and even with the same degree of purity it may be soft and fibrous, or hard and crystalline; therefore it is obvious that the young engineer should acquire a sound knowledge of its nature both chemically and practically, so as to enable him to obtain the quality he desires, and to know when he has secured it.

It would be easy to enlarge upon this interesting question of wrought-iron, but it may suffice to instance armour-plates, and rails, as cases where the best quality is required, but the quality, though best, must be different in kind; for armour-plates the iron can scarcely be too soft and fibrous, whilst for rails it can scarcely be too hard and crystalline, provided it is not so brittle as to be liable to fracture by use. Again, in some iron, such as the 'best Yorkshire,' the quality appears to improve with every additional operation in the manufacture, whilst the ordinary Welsh iron is almost destroyed by repeated manipulation. All these and many other matters connected with iron should therefore be known thoroughly and practically to the engineer.

In order to illustrate the necessity of the systematic study of the peculiarities of the metals called iron and steel, let me refer to the experiments of Mr. Eaton Hodgkinson, which first demonstrated that the average resistance of cast-iron to crushing was more than six times its tenacity, whilst the resistance of wrought-iron to crushing was only four-fifths of its tenacity, and it will be remembered that the mathematical investigations he founded upon these experiments first established on a satisfactory and reliable basis the degree and ratio of tensile and crushing force in cast and wrought-iron.

With respect to steel, it must be admitted, that before we can safely adopt it to any considerable extent for purposes of construction, it will be necessary to have a similar series of experiments and investigations specially made, but so promising a metal will amply repay all the trouble that may be bestowed upon it.

Steel.

Of timber a thorough knowledge should be acquired, as no material is otherwise more likely to deceive and to disappoint the engineer. Not only is great difference found in trees of the same general description, such as the numerous varieties of the pine, but the same kind of pine is a different quality of wood in different countries, and even in different soils and climate in the same country; and again the same tree is entirely changed by being 'bled' or having its sap withdrawn. The oaks of America, England, and the Continent are entirely different in their character, and oaks also differ in quality from each other in the same country, and so with numerous other woods used by the engineer. The strength, durability, and peculiarity of different kinds of timber, and the true value of artificially preserving them, should also be known and understood.

Timber.

I have selected these examples for the purpose of illustrating this important fact, that before an engineer can even commence the designs of his works he must have previously obtained a large amount of preliminary information regarding the nature of all the materials employed upon engineering works, so as to enable him to select for his intended structures those materials which will be on the whole the most suitable; having reference to efficiency, durability, and economy.

I will now proceed to the question of the kind and degree of knowledge which is required to enable a young engineer to proceed to the actual design of a public work of importance, such as a railway with its stone, brick, and iron struc- KNOWLEDGE REQUIRED tures, its earthworks, and its all-important permanent BY A CIVIL way; a railway station, a station roof; docks and their ENGINEER. appliances; waterworks, breakwaters, or a Great Eastern steamship.

Although it has become the practice in modern times for many civil engineers to be employed chiefly, or almost entirely, in some one branch of the profession, I desire to repeat my conviction that it is most important that the early preparation and subsequent study should be as extensive as possible, and should embrace every branch of professional practice, not only for the purpose of securing to a young engineer more numerous opportunities for his advancement, but also because sound knowledge and experience in all branches of engineering will greatly add to his efficiency and value in any special branch, in the same manner that a medical man will be more reliable in his practice on the eye and the ear if he possesses a sound practical and theoretical knowledge of every part of the human frame.

All classes of the profession, but especially the railway, the dock and harbour, and the waterworks engineer, must possess a knowledge of parliamentary proceedings, so as to be able to avoid all non-compliances with the Standing Orders of Parliament. To do this, it is true, is no easy matter, as the clauses are often drawn up with so little care and practical knowledge that neither engineers nor solicitors, nor the most experienced parliamentary agents, can understand what is intended.

On the subject of parliamentary proceedings generally, it may be taken for granted that all Committees desire to do justice to the cases which are brought before them, and that if they sometimes fail in their decisions, either as regards the interests of the public, or in arranging a fair settlement between antagonistic interests, it is not unfrequently due to the imperfect and crude manner in which cases are presented to them. I would therefore impress on all young engineers the importance, both to themselves and to their clients, of laying their cases before Committees in the most perfect manner possible, accompanied by full and correct information, carefully prepared and clearly worked out.

The professional knowledge required by the *railway engineer* commences with surveying of all kinds, the use of the theodolite, the Railway aneroid barometer, the level, the sextant, etc., and includes engineering. the surveys for preliminary and parliamentary purposes; and also working surveys of minute accuracy, on a large scale, from which engineering works may be set out with precision upon the ground.

The railway engineer must understand thoroughly the nature of earthworks of every kind, and the proper angles or slopes to be adopted for cuttings and embankments.

He must have the qualifications requisite to enable him to design bridges, viaducts, tunnels, and all other incidental works and buildings, in the best and most economical manner.

He must have a knowledge of the training of rivers, and of the effect of floods and drainage, in order that he may make accurate provision for the due discharge of water without wasting money on works unnecessarily large, or to avoid the risk of damage arising from making them insufficient.

He must be familiar with the various characters of permanent way, the best description of rail, sleeper, fastenings, and ballast, and with the different descriptions of switches, crossings, turntables, signals and telegraphs.

It is somewhat remarkable that, with all our experience, there should still remain a doubt amongst engineers as to the best kind of permanent way to be adopted even under similar circumstances. For although continental engineers have almost without exception adopted the flat-bottomed or 'Vignoles' form of rail, the ⊥ form of rail with equal top and bottom webs, and cast-iron chairs and wooden keys, is still largely used in this country.

A collection of facts with respect to the different descriptions of permanent way in use in this and other countries, with a view to a comparison of the advantages and disadvantages of each, would form a most interesting and important paper for the Institution, especially if it embraced all the recent experiments with reference to the use of steel rails.

The railway engineer should not be destitute of some knowledge of architecture and such a taste for those graceful outlines and simple appropriate details which should always characterise the works of an engineer, as to be able to avoid, on the one hand, the unnatural ornamentation which seems to have no connection with the structure, and, on the other hand, a disregard of either form, outline, or proportion.

But all such knowledge may fail if there be not a constant supervision and control over the quality of all the materials and workmanship employed upon the railway. And it is not too much to say that without the practical knowledge which is obtainable only by the actual performance of the duties of resident engineer, it is hopeless to expect that any engineer can be competent to undertake the responsibility of important works, or be fit to have large sums of money entrusted to him for expenditure. It is in the capacity of resident engineer that all previous preparation, both scholastic and professional, and all theoretical acquirements, become utilised and rendered of practical value, and it is only after much experience on different works of varied character, dimensions, and materials, and the acquisition of the power of discriminating between good and bad materials and workmanship, that a young student of engineering can claim to take rank as a 'Civil Engineer.'

The *dock* and *harbour engineer* requires the general and much of the special knowledge of the railway engineer, such as that which belongs to railways and tramways, and warehouses for goods; and to this he must add a vast amount of other special knowledge not required by the railway engineer.

Dock and harbour engineering.

For example, he must understand the laws which govern the ebb and flow of the tides, the rise and fall and time of high and low water; and he must have a knowledge of marine surveying, or the best means of ascertaining the set and speed of currents, and their tendency to increase depth of water by scour, or to diminish it by silting; he must also know, in the case of docks, what kind and extent of entrance accommodation to provide, whether the general plan should comprise only a simple lock, or be combined with a half-tide basin; whether single or double gates should be used; and whether it would be necessary to have a tidal basin or a recessed space, or both.

The nature of the trade to be accommodated in the proposed docks must also be carefully ascertained, in order to provide a proper proportion of quay space and water space, and proper width of quays, warehouses for bonding or for goods to be deposited, sheds for temporary protection, entrance for barges into warehouses from the docks, graving docks and workshops, with mechanical appliances for gates, sluices and pumping, and for shipping or discharging minerals or goods.

He may have to deal with solid foundation, and enjoy a facility of procuring suitable materials for construction, as at Liverpool; or he may have the bad foundations of Hull and other places, where alluvial silt of great depth has accumulated. It may be that good sound stone is too costly for the mass of his work, and that he must resort to brickwork, or rubble stonework, or concrete, or to a combination of all three; but in determining such questions it is impossible that anything but previous experience and habits of careful investigation will enable an engineer to arrive at the best decision. For it is not enough that his work should be solid, permanent, and safe, but it should be rendered so at the smallest possible cost.

The *dock* and *harbour engineer* is also required to report upon, and to construct, harbours of refuge, piers, landing stages, lighthouses, forts, canals and their appliances, river improvements, and many other hydraulic works; and in short, of this branch of engineering it may be truly said that questions are continually arising which require special study and mechanical invention to a greater extent than in almost any other branch of the profession.

Harbours of refuge being large and costly works, are necessarily few in number, and they are so slow in progress, and have generally been so often changed from their original object and design, that few engineering works have given less satisfaction either to the profession or the public; but we may hope, that if governments will accurately appreciate the objects they desire to obtain, and boldly grapple with the difficulties and cost of well-matured design, better and more useful works of this nature may be accomplished than have yet been undertaken.

The *waterworks* and *drainage engineer* must possess many of the qualifications of the railway and dock engineer, and especially those which concern earthwork and masonry; he must also be familiar with the means of obtaining information on the subject of rainfall in different localities, the methods of correctly gauging streams of every kind; the proportions of the rainfall available for his purposes after estimating for evaporation and waste, and the extent of the provision to be made for periods of dry weather, or for compensation to mill-owners and other interested parties.

Waterworks and drainage engineering.

He must be conversant with the proper mode of executing the works of reservoirs, conduits, weirs, tunnels, and aqueducts.

He must understand, by the aid of the chemist and his own experience, the nature of the impurities in water, and the best mode of diminishing them, whether mechanically, by subsidence and filtration, or otherwise.

To the waterworks engineer we must look for the solution of one of the great problems which the rapid increase of population is now forcing upon us, viz. a comprehensive system of conservancy of the flood waters of mountainous localities for the use of large cities and towns; and densely populated districts. We are completely outgrowing our present arrangements for water supply in the great majority of instances; and the convenience, comfort, and health of the public demand that such works when required shall be no longer postponed.

The initiative has been taken as to the question of a new source of water supply for London, in a pamphlet by a well-known authority in this branch of engineering, and sooner or later the subject must command public attention.

The waterworks engineer must also be competent to design and superintend works of sewerage, as well as of water supply, for large and small towns and localities; and his familiarity with waterworks will naturally aid him in this, as the problems for the discharge and pressure of fluids are identical in both cases.

The great sewerage works of London are now far advanced and have already produced beneficial results; the attention of other still neglected cities and towns has recently been called to this important subject by the loud and startling voice of a threatened return of cholera, and it is to be hoped that the proper authorities will perform their duty promptly and efficiently in this matter: but I cannot here refrain from calling attention to a gigantic evil which has been created by certain drainage and sewerage works already executed, and where the convenience and comfort of one set of people have been secured only by the infliction of a nuisance upon others. I allude to the discharge of collected sewage, without any attempt at purification or deodorisation, into streams of pure water.

It is remarkable that an injustice so great, and an evil so intolerable, should in any case have been permitted by Parliament, or by the general law of the land; but now that public attention has been fairly directed to the subject, let us hope that as soon as possible a remedy will be applied to the cases where mischief has already been done, and that care will be taken to prevent its recurrence.

It is no longer a matter of doubt that deodorisation or purification is quite practicable in every locality, and therefore no sewage should ever be permitted to be discharged into existing streams without this purification, or it should be carried out to the sea, and there discharged, as is now proposed for the north side of London.

The *mechanical engineer* deals with the most varied and numerous subjects of all the branches of engineering. They require that he should thoroughly understand the means of producing mechanical power, and of applying it to all the infinite variety of purposes for which it is now demanded. To this end he should be master of the laws of motion and rest, of power and speed, of heat and cold, of liquids and gases.

Mechanical engineering.

He must be familiar with the strength of materials under every variety of strain, the proper proportions of parts, and the friction of surfaces.

He must apply existing tools and contrive new ones for his work, and know how to direct power in the raising of weights, or for driving all fixed machinery, or in producing locomotion on land or water.

On railways he is responsible for the vast number of objects required in the machinery for erecting and repairing shops for the engines and carriages, for the pumping and other fixed engines, and especially for the locomotive engine itself, and for rolling and fixed plant generally.

In connection with docks he is required to design the machinery for opening and closing the dock gates, working sluices, emptying graving docks, or for working the cranes on the quays, or in the sheds and warehouses.

The mechanical engineer generally also executes the designs of the gas-engineer, even when he does not originate the work which is entrusted to him; and in this branch considerable chemical knowledge must be added to his mechanical qualifications.

For waterworks he designs and executes pumping engines, sluices, valves, stopcocks.

In the case of mines he supplies designs of the engines for pumping, drawing, winding, or ventilating; for locomotives above and below ground, as well as for the various mechanical appliances required in collieries, mines, and ironworks.

The adoption of the telegraph has been so astonishingly rapid

PRESIDENTIAL ADDRESS 381

and extensive, both by sea and land, and the purposes to which it
has been applied so important, that a considerable body of able and
accomplished engineers have devoted themselves almost exclusively
to the subject for the last few years, and have already created a new
branch of the profession, called *telegraphic engineering*; but to
be an accomplished telegraph engineer it is necessary first to be a
good mechanical engineer and then to add the special knowledge
of the electrician, and therefore I include telegraphic under the head
of mechanical engineers.

I think it may fairly be traced to the distinguished ability of that
class of mechanical engineers who have devoted themselves to telegraphic engineering, that already so much has been done in telegraphy.
Certainly no discussions have been more ably sustained in this Institution than those upon this subject.

Allied with the mechanical engineer is the *naval architect*, and
only a mechanical engineer could have constructed the vast steam-
ships of modern days. The ordinary timber-ship builder of old
would have been literally 'at sea' in the construction of modern
vessels, wherein the material is iron, and when the size of the vessel
requires scientific knowledge of form and resistance, of strains and
of strength, and when steam is the motive power. The demand
for large and swift vessels for ferries, for long voyages, for floating
batteries, and for iron-clad sea-going vessels, has of late been so
great that the construction of steam vessels has become a distinct
branch of engineering, under the name of naval architecture.

The *mining engineer* must possess much of the knowledge of the
railway and the mechanical engineer, and he must add to that
general knowledge much special knowledge of his own.
He must know how to sink shafts to the minerals if they Mining
engineering.
require to be extracted from beneath the surface (which
is usually the case), and how to divert or pump out the water he
meets with either in the shafts or the workings.

He must know how to excavate and bring to the surface minerals,
whether they be coal, copper, tin, lead, or iron, and to do this he
must construct subterranean railways, provide means of ventilation
by fans or furnaces, supply power to lift the extracted mineral to the
surface; and when brought there he must understand the further
requisite work, as the coal will probably require screening, or washing,
or manufacturing into coke, and the ore will require crushing, washing, or smelting, or possibly all three operations.

In all these cases, and many others, such as the collection of surface ironstone and other minerals, by railways and locomotive engines, and the working of lifts and inclined planes, the mining engineer has most important functions to perform, and has special machinery to adapt or invent; and relying on his judgment and skill alone, the investment of large sums of money for the development of the mineral wealth of this country is annually made.

I must not altogether omit a passing reference to the scientific talent which of late years has been devoted to Artillery—its weapons of attack and works of defence; and I think we may fairly claim that it is mainly due to some of the able members of this Institution that this *art* has been placed on a new and vastly improved basis, and that as a consequence a new branch of the profession has been actually created—Artillery Engineering.

Artillery engineering.

Having now enumerated in some detail the various descriptions of work which engineers are called upon to carry out, I will next proceed to point out the kind of preparation which, in my opinion, is requisite to enable them to perform their work in a proper manner.

Preparation required by a Civil Engineer.

I am aware of the difficulty of the task, and of the wide difference of opinion which exists on the subject, but I feel unable to resist the opportunity of bringing this question under the consideration of the Institution, because I feel convinced that at no period in the history of the profession has it been so important as at the present time. Those who may not be disposed to coincide in my views may at least be led by the description of them to throw new light on a subject which is of vital consequence.

We of the passing generation have had to acquire our professional knowledge as we best could, often not until it was wanted for immediate use, generally in haste and precariously, and merely to fulfil the purpose of the hour, and therefore it is that we earnestly desire for the rising generation those better opportunities and that more systematic training for which in our time no provision had been made, because it was not then so imperatively required.

The preparation and training for the civil engineer may be shortly described as follows:

1. General instruction, or a liberal education.
2. Special education as a preparation for technical knowledge.
3. Technical knowledge.
4. Preparation for conducting practical works.

All this preparation and training will have to be acquired at some time or other, and in some order or other, and it is known that in the cases of some successful persons of great perseverance, they have been acquired in a very remarkable order; but at the present time, and with all our modern opportunities, there is no reason why they should not be learned in the most convenient and methodical manner.

I will begin by supposing a boy of fourteen, in whom his parents have discovered a mechanical bias, who has made good progress in his general education, and especially in arithmetic, is of strong constitution, and possessed of considerable energy and perseverance : and unless a boy possesses these tendencies and qualifications it is quite useless to destine him for an engineer.

Taking the boy of fourteen, however, who possesses the requisite qualifications, and with a determination on his own and his parents' part that he shall be made an engineer, the period from fourteen to eighteen should be devoted to the special education required by an engineer, during which mathematics, natural philosophy, land surveying and levelling, drawing, chemistry, mineralogy, geology, strength of materials, mechanical motions, and the principles of hydraulics should be thoroughly mastered.

To accomplish these studies, and, in addition, to make considerable progress in the living languages, French and German especially, it will be necessary to sacrifice to some small extent his classical studies and pure mathematics, and it is, in fact, the partial omission of these studies, and the prominence of those I have enumerated, which constitutes a 'special education.'

If from fourteen to eighteen the boy has made all the progress in these studies which can be reasonably expected from fair abilities and more than average perseverance, the next step is of great importance, and is one respecting which some difference of opinion will exist.

At eighteen a boy if duly prepared may either be at once placed in the office of a civil engineer for a period of four or five years' pupilage, or he may be placed in a mechanical workshop, or he may be sent to one of our great universities; and any one of these courses may be the best under particular circumstances, such as local convenience, or as the social position of parents may dictate.

It cannot be doubted that a period of twelve to twenty-four months may be very profitably spent in manufacturing works, before passing into a civil engineer's office ; but in that case the greatest possible care must be taken that the works selected are adapted in themselves to impart the desired information; and that proper organisation exists for carrying out strict office discipline, regularity of attendance and due diligence; and that assistance be given systematically to the pupil to enable him to obtain all the advantage possible from his stay at the works.

It is of the greatest importance to the future success of the engineer that during his professional preparation he should continue. his studies of mathematics and scientific works relating to his profession, and also of modern languages.

In the case of its being intended to send the boy to Cambridge or Oxford, it is indispensable that all preliminary professional work, such as practical knowledge of mechanics, mechanical drawing, surveying and levelling, should be mastered before going to the university, because it can scarcely be expected that he will submit to the drudgery of learning them after his return from a three years' university course, then at the age of say twenty-two. Probably the best plan will be to take him away from his scholastic studies somewhat earlier than eighteen, if it be intended that he should go to the university, and to take especial pains to make him accomplished in the preliminary work of the draughtsman, the surveyor, and the mechanic; so that when he has taken his degree and enters as a pupil in a civil engineer's office he will at once commence useful and interesting employment, and will not require more than three years' pupilage.

If arrangements can be so made, and assuming a boy has worked well at school with his general studies, and subsequently with his special studies; and if from the age of seventeen or eighteen he does justice to his opportunities in a good workshop, keeps up his knowledge of modern languages, proceeds to Cambridge or Oxford, taking a good degree, and afterwards completes his studies as a pupil with a civil engineer; probably such a course would constitute the best possible preparation and training which could be obtained: but at the same time it cannot be doubted that it is a somewhat hazardous combination, and can only be successful with great determination on the part of the pupil to keep his future career always in view, and to prepare for it accordingly, as well before going to the university, and during his college career, as after he leaves it.

With respect to the special preparation of young men between the ages of fourteen and seventeen or eighteen, several of the largest and best proprietary schools and colleges in this country have special classes and departments for the study of the applied sciences; and thence well-prepared pupils are annually sent out to commence their career with engineers, architects, and surveyors; but still the character of this special preparation, in its theoretical branches, is not considered quite equal to that of France or Germany for the civil engineer.

It is true that nearly all continental nations have an advantage over this country in the power which the nature of their government gives them of concentrating, in one recognised official school for the preparation of civil engineers, all the best available talent of their country.

This plan does not exist in our country, and on the whole we rejoice that it does not; neither does the inducement of government employment form the chief stimulus to our exertions, for which we are also thankful: but at the same time no good reason can exist why the opportunities of acquiring theoretical preparation in this country should be inferior to those of the Continent; and I have the confident hope, from the anxiety which is now manifested to increase the ranks of our profession, and the desire to have the best possible preparation for it, that even in the theoretical branches we shall shortly have to acknowledge no inferiority to any other nation. In the practical branches we are admittedly superior.

In drawing attention, however, to a comparison between our own and other countries, let me be guarded against the possibility of being understood to suggest that this theoretical equality ought to be obtained by any sacrifice whatever of our undoubted great practical knowledge. Indeed, on the contrary, I think that attention to the greater opportunities which young engineers in this country enjoy, by reason of the number and character of our new public works, than is attainable in other countries, should be constantly encouraged to the utmost possible extent, and that our superiority as practical engineers should be ever maintained.

We will now suppose that the general education and the special instruction have been completed, the short probationary pupilage in workshops has been gone through, languages and mathematics kept up and improved, the university course in certain cases completed, and the period has arrived for entering a civil engineer's office.

In selecting such office for a pupil it is important that it should be well organised and not be too large; that the engineer should be a comparatively young and rising man, and be accustomed to take pupils; but these should be few in number, and bear some pro-

2 c

portion to the number and extent of the works in usual course of construction under the engineer's direction.

It is not necessary to follow the pupil, when once the engineer's office is entered, with any detailed advice, because he is no longer a boy, unable to appreciate his position and duty. We assume that he has been highly educated and carefully trained, and knows well that his future success or failure will depend on the degree of diligence with which he avails himself of the opportunities of acquiring knowledge during his pupilage.

The work in the office and in the field should be done to the best of his ability, and after the pupil has become a skilful draughtsman, and is capable of taking out quantities of engineering works, and preparing detailed estimates, methodically arranged, he will then probably proceed to work out details of designs, and make calculations of strengths and strains, and thus become of real value in the office, at the same time making substantial progress and rapid improvement for himself.

He should avail himself of every opportunity of mastering the purpose and the principles of construction of the work brought to his notice, both in the office and in execution; and he should ascertain the cost price of all the materials and workmanship employed, separating the items into every minute detail.

The information which, amongst much beside, should be thus obtained during pupilage, and which is necessary to constitute a sound engineer, is—

1. A fair knowledge of the most fitting material to use for any given work, under any given circumstances.

2. The power of designing any ordinary work with a maximum of strength and a minimum of material and labour.

3. A knowledge of the means of ascertaining the cost price of any ordinary engineering work.

The information or knowledge included in this brief enumeration may be called practical knowledge, and it cannot be too often urged upon young engineers that theory and practice must always go together, hand in hand, step by step; and that they are not only not inconsistent or conflicting, but that they are necessarily united, and must both be fully developed in the same person before he can become a properly qualified 'Civil Engineer.'

The period of pupilage should be from three to five years, depending on the circumstances which have been previously indicated, and, in addition to his attention to the office, and to outdoor works, it will be well, while keeping up his preparatory studies, especially in mathematics, that he should improve his acquaintance with the French and German languages, and keep up his knowledge of their engineering literature, and also avail himself professionally and personally of the advantages offered by this Institution.

In the case of the mechanical engineer, however, it will be seen that although all scholastic and scientific training should be the same as that previously described for all other branches, the period of pupilage of the mechanical engineer must necessarily be passed chiefly in large workshops or manufacturing establishments.

I propose now to consider in what manner this Institution can be made available in the preparation of the young THE engineer, and more useful to the profession generally; INSTITUTION. and as a first step allow me, very briefly, to trace its history and refer to its present prosperity.

It will be remembered that the Institution of Civil Engineers was established on January 2nd, 1818, and that Telford was formally installed President on March 21st, 1820.

The origin of the Institution was very humble.

About the year 1816 Mr. Henry Robinson Palmer, who was then articled to Mr. Bryan Donkin, suggested to Mr. Joshua Field the idea of forming a Society of young engineers for their mutual improvement in mechanical and engineering science. The earliest members were Mr. Palmer, Mr. Field, and Mr. Nicholas Maudsley, to whom were shortly added Mr. James Jones, Mr. Charles Collinge, and Mr. James Ashwell.

When the Society was constituted, on January 2nd, 1818, these six young men were joined by two others, Mr. Thomas Maudsley and Mr. John T. ·Lethbridge, with Mr. James Jones as Secretary, and during the remainder of that year there was no increase in the number of the members, and the only additions were three new members in 1819.

In the following year, 1820, when Telford became President, there were thirty-two elections.

At the end of 1822, when the Institution had been established for five years, there had been fifty-four elections.

Telford's name gave a great impulse to the progress of the Institution, which grew rapidly in importance under his fostering hand, so that at the tenth year of its existence—at the close of 1827, there had been a total of 158 elections, and by June 3rd, 1828, when the charter of incorporation under the great seal was obtained, the number amounted to 185 members.

Telford continued to be the President until his decease occurred, which took place on September 2nd, 1834, and at that time the *actual* number of members on the books (as distinct from the number elected) was 200.

Mr. James Walker, the second President, was elected to that post on January 20th, 1835; and after occupying the chair for ten years, he declined to allow himself to be again put in nomination, in consequence of a strong expression of opinion from several influential members that a shorter period for the term of the office of President had become necessary.

Accordingly on January 27th, 1845, Sir John Rennie was elected President and served for three years.

Since then the chair has been successively filled by Joshua Field, Sir William Cubitt, James Meadows Rendel, James Simpson, Robert Stephenson, M.P., Joseph Locke, M.P., George Parker Bidder, John Hawkshaw, and John Robinson McClean, each of whom has served for *two* years, the maximum time now allowed by the by-laws.

It should be mentioned that in the ordinary course of rotation Isambard Kingdom Brunel would have succeeded Robert Stephenson, but Brunel requested that he might not then be put in nomination, owing to ill-health and the pressure of professional duties, and unhappily his early subsequent decease deprived the Society of any future opportunity of electing him. It must always be a subject of regret to the profession, that in the annals of the Institution a member so gifted and accomplished should not appear on their list of Presidents.

At the close of 1836, when the Institution had existed nineteen years, the number of members of all classes who had been elected was 369, and the number of those still remaining on the books was 252, or about five-sevenths of those elected.

At the close of 1860 these numbers were 1535 and 930 respectively, from which it appears that three-fifths of all those elected still belonged to the Institution, being a decline of only *one-seventh* in the relative proportions after a further existence of twenty-five years.

The average annual effective increase of members and associates during the ten years from 1840 to 1850 was 25, and from 1850 to 1860 it was 27, the actual increase in 1859 and 1860 being 37 in each year. In 1861 it was 20, and in 1862 the number was 57.

The numbers of members of all classes on the books on November 30th, 1865, were :—

Honorary Members	.	.	.	20
Members	.	.	.	486
Associates	.	.	.	689
Graduates	.	.	.	8
Total	.	.	.	1203

or an effective increase in one year of 108 members of all classes.

It will thus be seen that a steady annual increase has been the characteristic of the Institution from its commencement, and it may be noted in this, the forty-eighth year of its existence, that, when it had been established twenty-four years, the number of members was almost exactly one-half of the present number.

The experience of the last few sessions shows us clearly that we may expect the future rate of increase to be at least equal to the past, and the attendances on the Tuesday evening dis- *Attendance.* cussions show that the interest attached to the proceedings of the Institution increases in at least an equal proportion with the augmentation of the numbers.

It is now not uncommon to find our meeting-hall inconveniently crowded, and occasionally it is altogether inadequate to accommodate the numbers who desire to be present; and many persons who, from the public interest attached to some of the subjects, desire to hear or to take part in the discussions, are now prevented by our restricted accommodation from doing so.

For some years in the early history of the Institution it was a work of considerable difficulty to keep the disbursements within the receipts, and except for the admirable management of our *Finance.* late Secretary and now Honorary Secretary, Mr. Manby, it is hard to know what difficulties we might not have experienced. It was not until its income became sufficiently increased by the liberal donations of the council and other members, by trust-moneys and bequests, and by the increase in its numbers, that the Institution was in a financial position to give increased accommodation and assistance to its members.

It may be stated that during the last ten years the average increase on the receipts has been forty per cent., whilst the increase in the disbursements has been only twenty per cent.; and that the present amount of the realised property of the Institution may be safely taken at £25,000.

It will have been observed that considerable improvements have been made in the library of the Institution, and in its arrangements and facilities; and no doubt the Council and Secretary will continue to give this important department their earnest attention, and we may reasonably expect that both the contents of the library and its accessibility will be still further increased.

Library.

It is, however, somewhat remarkable that a greater number of members do not avail themselves of the additional opportunities of reference to the library which have been afforded them, and this brings me at once to the consideration of the important question of the manner in which this Institution may be made more useful to its members.

Additional facilities.

The state of the finances, as we have already seen, will prudently permit the expenditure of a larger annual sum than we now disburse, and therefore we are at full liberty financially to consider the question of additional accommodation for the members, and I believe the library of the Institution would be far more valuable if an arrangement could be made by which it might be kept open in the evenings for a certain number of days in the week, say until nine or ten o'clock. I have ascertained that no practical obstacle to this extension of use exists, and that the additional expense would not be considerable.

Most of the members of the Institution are necessarily engaged in their ordinary daily professional duties during the only hours when the library is at present available to them, and it is obvious that it is only in the case of a special reference being required, or for some statistical purpose, that the library can be useful to members generally under the present arrangement.

I can say from my own experience that I should have felt it a great boon, as a young man, to have had the opportunity of spending an occasional evening in the library, and of reading and consulting the rich record of professional learning and experience now collected there, and therefore I throw out this hint respecting the extension of the hours for reading.

Another step might probably be taken with great advantage to students and engineers generally, viz. the systematic collection of good working drawings, specifications, and contracts for important works in progress or completed, with facility for reference to them in the library, and permission to make tracings or copies.

There can be little doubt that engineers in large practice would permit copies to be taken of their working drawings and specifications for this purpose, and in addition to this assistance with respect to drawings, it would not be difficult to obtain permission for the inspection of the works themselves, during their execution, so that young engineers might have the opportunity, especially during the summer months, of seeing works as they are carried out, and comparing them with the drawings and specifications to which they have had access in the library.

I would also venture to suggest that, in addition to the greater advantage which may be conferred on those using the library by extended time of access to it, and to the collection of working drawings and specifications, with arrangements for inspection of practical works, a limited number of lectures would be very valuable if given by members who were especially conversant with any given subject, on other evenings than those of the ordinary meetings during the session of the Institution.

I now approach a question in connection with the Institution and its functions upon which, in common with the profes- NEW BUILD-sion generally, I confess I feel very strongly, and that is, ING. the necessity of providing as soon as possible a building more commodious and more convenient than that which we now possess.

Our rapidly increasing numbers have already reached the point when, as I have previously stated, the theatre in which we are now assembled is admittedly insufficient for the accommodation of those who wish to attend our discussions; and, in addition to inadequate space, there are conditions inseparably attached to the present building which prevent this room being properly ventilated and rendered comfortable.

The other rooms of this building are also totally inadequate for the ordinary purposes for which they are required, and on the evenings of our annual *conversazione* especially, the crowding and discomfort are such as to repel many of our best friends from venturing to be present with us.

With a proper building and well-arranged rooms, we shall also be able to have many objects of professional interest for our inspection and study, of which we are at present deprived—such as models of work and machinery, new articles or new combinations, or, possibly, even a good museum.

I hope, however, we shall shortly be in a position to consider a distinct proposal for a new building, worthy of the present position and the future requirements of the Institution.

Having now frankly brought before the Institution some of the more important matters which appear calculated to influence the future of the members of our profession, permit me to say, in conclusion, that I am not sanguine enough to expect that I shall accomplish more in this Address than direct the thoughts and attention of my professional brethren to the subject, and induce others more able than myself to take it up.

CONCLUSION.

It cannot be doubted that the rapidly increasing prominence and importance of our profession imposes upon us grave responsibility and the duty of vigilant watchfulness, so that the character of our members, and the success of our works, may be all that greater knowledge, wider experience, and more cultivated taste ought to make them, and that every new work of importance may be better than that which has preceded it, and remain as a monument of progress of which all may be proud.

It is not now sufficient that an engineering work should be durable and free from failure, but, with our present means of study and of knowledge, it will be expected that our works should display in a satisfactory degree the qualities of fitness, economy, and taste, in addition to that of durability.

With deeper study and more complete preparation, the love of our profession and pride in its noble works will become greater and greater in its students, and lead to that intense devotion and application which history teaches us has alone produced the greatest works in art and science; and we cannot doubt that far greater triumphs remain to be, and will be, achieved, by those whom I now see before me, than have yet been realised by either ancient or modern engineers.

Amidst all the excitement of our professional avocations, however, let us constantly bear in mind, and endeavour to imitate, the example of the distinguished men who have been removed from amongst us during the last few years in the happy manner in which they succeeded in combining personal friendship with professional rivalry, and in their never-failing interest in the prosperity and usefulness of this Institution.

INDEX

Milton Keynes UK
Ingram Content Group UK Ltd.
UKHW032320161024
449665UK00001B/37